William Robert Ogilvie-Grant

A Hand-Book to the Game-Birds

Vol. I: Sand-Grouse, Partridges, Pheasants

William Robert Ogilvie-Grant

A Hand-Book to the Game-Birds
Vol. I: Sand-Grouse, Partridges, Pheasants

ISBN/EAN: 9783744735421

Printed in Europe, USA, Canada, Australia, Japan

Cover: Foto ©berggeist007 / pixelio.de

More available books at **www.hansebooks.com**

A HAND-BOOK
TO THE
GAME-BIRDS.

BY

W. R. OGILVIE-GRANT,

ZOOLOGICAL DEPARTMENT, BRITISH MUSEUM.

VOL. I.

SAND-GROUSE, PARTRIDGES, PHEASANTS.

LONDON:
W. H. ALLEN & CO., LIMITED,
13, WATERLOO PLACE, S.W.
1895.

PREFACE.

The name of my colleague, Mr. Ogilvie-Grant, is now so well known as an authority on the Game-Birds that very few remarks are necessary to introduce him to my readers. The work is founded on his volume of the "Catalogue of Birds in the British Museum," where the student will find detailed the material, on which he has grounded the present monographic review. The aim of the Author has been to provide such a "Hand-book" as may be useful to sportsmen in every part of the world, and the present volume will prove of service to travellers in Africa, as it gives a diagnosis, whereby every species of Francolin, known up to the present time, may be distinguished.

The second volume will deal with the Pheasants, American Partridges, Megapodes, Curassows, and Hemipodes, in the same concise manner, and will, I believe, be found of equal service to the sportsman and naturalist.

R. BOWDLER SHARPE.

AUTHOR'S PREFACE.

IN preparing the present volume, which includes the first half of the species commonly termed "Game-Birds," my great aim has been to treat the subject in such a way that this little book may not only be useful as a scientific work of reference, but also as a handy book for sportsmen and field naturalists. With its aid, they should be able not only to identify the birds they shoot, with as little trouble as possible, but also to find out what is known concerning the life-history of each species.

References are, in every case, given to the more important works, especially those in which good figures of the birds are to be found.

The descriptions of the adult male and female have been made as short as possible, only the distinguishing characters being given, while the more important points are printed in italics; and it is believed that, in every case, the descriptions will be found quite sufficient to enable those who have no previous knowledge of this group, to identify any species of Game-Bird they may chance to meet with.

In such birds as the Seesee Partridges, and in some of the closely allied species of Kalij and Koklass Pheasants, the females so closely resemble one another, that it has been found impossible to give characters by which they may be distinguished one from another. In such cases the best guide to identification is the *locality* (if that is known) in which the individual bird was obtained.

My endeavour has been, as far as possible, to give the description, &c., in the plainest language, devoid of scientific phraseology, but should the reader ever be in doubt as to which part of the bird is referred to, he has only to turn to the diagram (of a Francolin) given at the beginning of the book (p. xvi.), which will clearly explain the terms employed in the description.

I have to acknowledge the great assistance I have received from the works of Captain Bendire on the "Life History of North American Birds," and the notes published by Mr. A. O. Hume, C.B., in the "Game Birds of India." On the Grouse and Ptarmigan I have also derived much useful information from the "Shooting Sketches" of Mr. J. G. Millais.

W. R. OGILVIE-GRANT.

SYSTEMATIC INDEX.

	PAGE
ORDER PTEROCLETES.	1
FAMILY I. PTEROCLIDÆ.	3
I. SYRRHAPTES, Illig.	3
1. paradoxus (Pall.).	3
2. tibetanus, Gould.	6
II. PTEROCLURUS, Bp.	7
1. alchatus (L.).	8
a. pyrenaicus (Seeb.).	10
2. namaquus (Gm.).	11
3. exustus (Temm.).	12
4. senegallus (L).	14
III. PTEROCLES, Temm.	15
1. arenarius (Pall.).	15
2. decoratus, Cab.	16
3. variegatus, Smith.	17
4. coronatus, Licht.	18
5. gutturalis, Smith.	19
6. personatus, Gould.	20
7. lichtensteini, Temm.	20
8. bicinctus, Temm.	21
9. fasciatus (Scop.).	22
10. quadricinctus, Temm.	24
ORDER GALLINÆ.	25
FAMILY I. TETRAONIDÆ.	26
I. LAGOPUS, Briss.	26
1. scoticus (Lath.).	27
2. lagopus (L.).	36
3. mutus (Montin).	38
4. rupestris (Gm.).	42
5. hyperboreus, Sundev.	43
6. leucurus, Swains. & Rich.	44

		PAGE
II. LYRURUS, Swains.		45
1. tetrix (L.).		45
2. mlokosiewiczi (Tacz.).		48
III. TETRAO, L.		49
1. urogallus, L.		49
a. uralensis, Nazarov.		52
2. parvirostris, Bp.		53
3. kamtschaticus, Kittl.		54
IV. CANACHITES, Stejn.		54
1. canadensis (L.).		54
2. franklini, Dougl.		56
V. FALCIPENNIS, Elliot.		57
1. falcipennis (Hartl.).		57
VI. DENDRAGAPUS, Elliot.		58
1. obscurus (Say).		58
a. fuliginosus (Ridgw.).		60
2. richardsoni (Dougl.).		61
VII. TYMPANUCHUS, Gloger.		61
1. americanus (Reichenb.).		62
2. cupido (L.).		65
3. pallidicinctus (Ridgw.).		65
VIII. CENTROCERCUS, Swains.		66
1. urophasianus (Bp.).		66
IX. PEDIŒCETES, Baird.		68
1. phasianellus (L.).		68
2. columbianus (Ord.).		69
X. BONASA, Steph.		71
1. umbellus (L.).		71
XI. TETRASTES, Keys. u. Blas.		74
1. bonasia (L.).		74
2. grisciventris, Menzb.		77
3. severtzovi, Prjev.		77
FAMILY II. PHASIANIDÆ.		78
SUB-FAMILY I. PERDICINÆ.		79
I. LERWA, Hodgs.		79
1. lerwa (Hodgs.).		80

	PAGE
II. TETRAOPHASIS, Elliot.	81
1. obscurus (Verr.).	81
2. széchenyii, Madar.	83
III. TETRAOGALLUS, J. E. Gray.	83
1. tibetanus, Gould.	84
2. henrici, Oust.	85
3. altaicus (Gebler).	86
4. himalayensis, Gray.	86
5. caspius (Gm.).	89
6. caucasicus (Pall.).	90
IV. CACCABIS, Kaup.	90
1. saxatilis (W. and M.).	90
a. chukar (J. E. Gray).	91
2. magna, Prjev.	95
3. rufa (L.).	96
4. petrosa (Gm.).	97
5. melanocephala (Rüpp.).	98
V. AMMOPERDIX, Gould.	99
1. bonhami (Fraser).	99
2. heyi (Temm.).	101
VI. FRANCOLINUS, Steph.	101
1. francolinus (L.).	103
2. pictus (J. and S.).	106
3. chinensis (Osbeck).	107
4. lathami, Hartl.	108
5. pondicerianus (Gm.).	108
6. coqui (Smith).	111
7. hubbardi, Ogilvie-Grant.	112
8. schlegelii, Heugl.	112
9. streptophorus, Ogilvie-Grant.	112
10. sephæna (Smith).	113
11. granti, Hartl.	114
12. kirki, Hartl.	114
13. spilogaster, Salvad.	114
14. albigularis, Gray.	115
15. spilolæmus, Gray.	115
16. gutturalis (Rüpp.).	116
17. uluensis, Ogilvie-Grant.	117

FRANCOLINUS—(continued).

		PAGE
18. africanus, Steph.	...	117
19. finschi, Bocage.	...	118
20. castaneicollis, Salvad.	...	118
21. levaillanti (Valenç.).	...	119
22. gariepensis, Smith.	...	120
23. jugularis, Büttik.	...	121
24. shelleyi, Ogilvie-Grant.	...	121
25. elgonensis, Ogilvie-Grant.	...	122
26. gularis (Temm.).	...	122
27. adspersus, Waterh.	...	124
28. griseostriatus, Ogilvie-Grant.	...	125
29. bicalcaratus (L.).	...	126
30. clappertoni, Childr.	...	126
31. gedgii, Ogilvie-Grant.	...	127
32. hartlaubi, Bocage.	...	127
32a. dybowskii, Oust.	...	287
33. icterorhynchus, Heugl.	...	128
34. sharpii, Ogilvie-Grant.	...	128
35. capensis (Gm.).	...	129
36. natalensis, Smith.	...	130
37. hildebrandti, Cab.	...	131
38. johnstoni, Shelley.	...	132
39. fischeri, Reichenow.	...	132
40. squamatus, Cass.	...	132
41. schuetti, Cab.	...	133
42. ahantensis, Temm.	...	133
43. jacksoni, Ogilvie-Grant.	...	134
44. erckeli, Rüpp.	...	135

VII. PTERNISTES, Wagler. ... 135

1. nudicollis (Bodd.).	...	136
2. humboldti (Peters).	...	136
3. afer (P. L. S. Müll.).	...	137
4. cranchi (Leach).	...	138
5. boehmi, Reichenow.	...	138
6. swainsoni (Smith).	...	139
7. rufopictus, Reichenow.	...	140
8. leucoscepus, Gray.	...	140
9. infuscatus, Cab.	...	141

SYSTEMATIC INDEX.

	PAGE
VIII. RHIZOTHERA, Gray.	141
1. longirostris (Temm.).	142
2. dulitensis, Ogilvie-Grant.	142
IX. PERDIX, Briss.	143
1. perdix (L.).	143
a. damascena, Briss.	148
2. daurica (Pall.).	149
3. hodgsoniæ (Hodgs.).	150
4. sifanica, Prjev.	151
X. MARGAROPERDIX, Reichenb.	151
1. madagascariensis (Scop.).	152
XI. PERDICULA, Hodgs.	153
1. asiatica (Lath.).	153
2. argoondah (Sykes).	155
XII. MICROPERDIX, Gould.	156
1. erythrorhyncha (Sykes).	156
2. blewitti, Hume.	158
3. manipurensis (Hume).	159
XIII. ARBORICOLA, Hodgs.	160
1. torqueola (Valenç.).	160
2. atrigularis, Blyth.	163
3. ardens, Styan.	164
4. crudigularis (Swinh.).	164
5. intermedia, Blyth.	165
6. rufigularis, Blyth.	165
7. gingica (Gm.).	166
8. mandellii (Hume).	167
9. javanica (Gm.).	167
10. rubrirostris (Salvad.).	168
11. brunneipectus, Tick.	169
12. hyperythra (Sharpe).	170
13. erythrophrys (Sharpe).	171
14. orientalis (Horsf.).	171
15. sumatrana (Ogilvie-Grant).	172
XIV. TROPICOPERDIX, Blyth.	172
1. chloropus, Blyth.	172
2. charltoni (Eyton).	173

	PAGE
XV. HÆMATORTYX, Sharpe.	174
1. sanguiniceps, Sharpe.	174
XVI. CALOPERDIX, Blyth.	175
1. oculea (Temm.).	175
a. sumatrana, Ogilvie-Grant.	176
b. borneensis, Ogilvie-Grant.	176
XVII. ROLLULUS, Bonn.	177
1. roulroul (Scop.).	177
XVIII. MELANOPERDIX, Jerd.	178
1. nigra (Vig.).	179
XIX. COTURNIX, Bonn.	179
1. coturnix (L.).	180
a. capensis, Licht.	183
2. japonica, Temm. and Schl.	184
3. coromandelica (Gm.).	185
4. delegorguei, Deleg.	187
5. pectoralis, Gould.	187
6. novæ-zealandiæ, Q. and G.	188
XX. SYNŒCUS, Gould.	190
1. australis (Temm.).	190
2. raalteni (Müll. and Schl.).	192
3. plumbeus, Salvad.	192
XXI. EXCALFACTORIA, Bp.	193
1. chinensis (L.).	193
2. lineata (Scop.).	196
3. lepida, Hartl.	197
4. adansonii (Verr.).	197
SUB-FAMILY II. PHASIANINÆ.	199
I. PTILOPACHYS, Swains.	199
1. fuscus (V.).	199
II. BAMBUSICOLA, Gould.	202
1. fytchii, Anderson.	202
2. thoracica (Temm.).	203
3. sonorivox, Gould.	204
III. GALLOPERDIX, Blyth.	205
1. spadicea (Gm.).	206

GALLOPERDIX—(continued).	
2. lunulata (Valenç.)	208
3. bicalcarata (Penn.)	210
IV. OPHRYSIA, Bp.	212
1. superciliosa (Gray)	213
V. ITHAGENES, Wagl.	214
1. cruentus (Hardw.)	215
2. geoffroyi, Verr.	218
3. sinensis, David.	219
VI. TRAGOPAN, Cuv.	220
1. satyra (L.)	220
2. melanocephalum (J. E. Gray)	224
3. temmincki (J. E. Gray)	227
4. blythii (Jerd.)	228
5. caboti (Gould)	229
VII. LOPHOPHORUS, Temm.	230
1. refulgens, Temm.	231
a. mantoui, Oust.	236
b. obscurus, Oust.	236
2. impeyanus (Lath.)	237
3. l'huysii, Verr. and Geoffr.	238
4. sclateri, Jerd.	240
VIII. ACOMUS, Reichenb.	240
1. erythrophthalmus (Raffl.)	241
2. pyronotus (Gray)	242
3. inornatus, Salvad.	242
IX. LOPHURA, Flem.	243
1. rufa (Raffl.)	244
2. ignita (Shaw)	246
3. diardi (Bp.)	247
X. LOBIOPHASIS, Sharpe.	248
1. bulweri, Sharpe.	249
XI. CROSSOPTILON, Hodgs.	251
1. tibetanum, Hodgs.	252
2. leucurum, Seeb.	253
3. manchuricum, Swinh.	254

		PAGE
CROSSOPTILON—(*continued*).		
4. auritum (Pall.), ...	... ...	255
5. harmani, Elwes.	... ...	257
XII. GENNÆUS, Wagl. ...	... ...	258
1. albocristatus, Vig.	... ...	258
2. leucomelanus (Lath.)	... ...	262
3. melanonotus (Blyth)	... ...	263
4. horsfieldi (Gray).	... ...	269
a. cuvieri (Temm.).	... ...	271
b. davisoni, Ogilvie-Grant.	... ...	271
5. lineatus (Vig.). ...	... ...	272
a. oatesi, Ogilvie-Grant.	... ...	276
6. andersoni (Elliot).	... ...	276
7. nycthemerus (L.).	... ...	277
8. swinhoii, Gould.	... ...	278
XIII. PUCRASIA, Gray.	... ...	280
1. macrolopha (Less.).	... ...	281
a. biddulphi, Marshall.	... ...	284
2. nipalensis, Gould.	... ...	284
3. castanea, Gould.	... ...	285
4. meyeri, Madar.	... ...	285
5. xanthospila, Gray.	... ...	286
6. darwini, Swinhoe.	... ...	287

LIST OF PLATES.

	TO FACE PAGE
I.—Pallas' Three-toed Sand-Grouse.	4
II.—Feathers of Scotch Grouse.	27
III.— ,, ,, ,,	31
IV.—Rock Ptarmigan.	43
V.—Ural Capercailzie.	52
VI.—Sage Grouse.	67
VII.—Northern Sharp-tailed Grouse.	68
VIII.—Ruffed Grouse.	71
IX.—Altai Snow-Cock.	86
X.—Spanish Red-legged Partridge.	96
XI.—Common Francolin.	103
XII.—Mountain Partridge.	147
XIII.—Manipur Painted Bush-Quail.	159
XIV.—Mandelli's Tree-Partridge.	167
XV.—Red-crested Wood-Partridge	177
XVI.—Japanese Quail.	185
XVII.—Blood Pheasant.	215
XVIII.—Cabot's Tragopan.	229
XIX.—Chamba Moonal Pheasant.	Frontispiece
XX.—Bulwer's Wattled Pheasant.	249
XXI.—Koklass Pheasant.	281

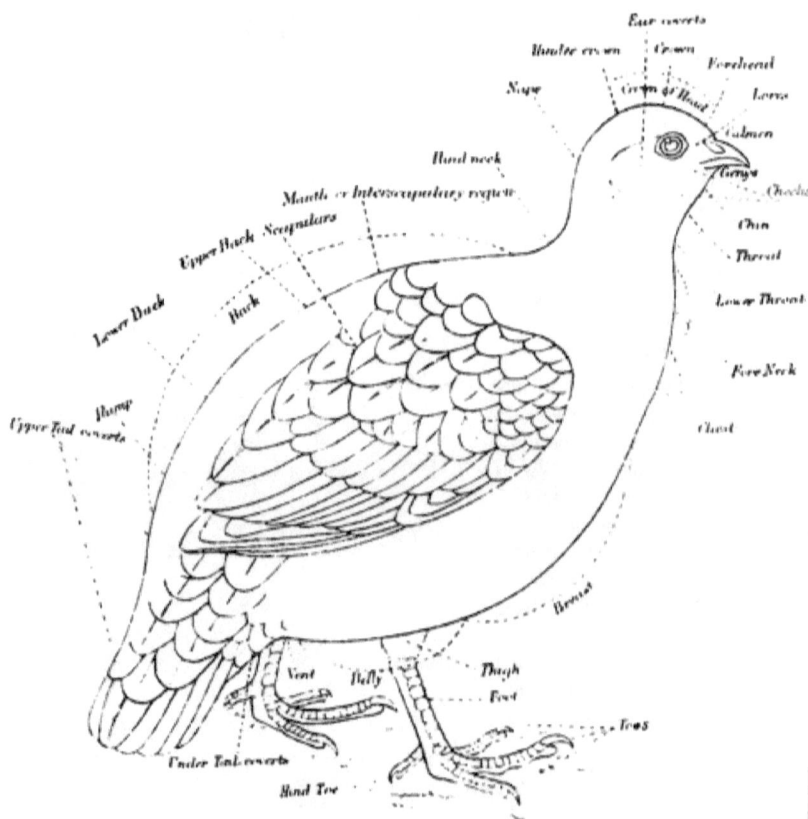

Outline figure of a Francolin to illustrate the nomenclature employed in describing a bird.

GAME-BIRDS.

THE SAND-GROUSE. ORDER PTEROCLETES.

THE sixteen species comprising this small Order, intermediate in its affinities between the Pigeons (*Columbæ*) and the True Game-Birds (*Gallinæ*), are all included in one family, *Pteroclidæ*. Their general structure presents many striking Columbine characters, as in the vocal organs, pterygoid bones, and the

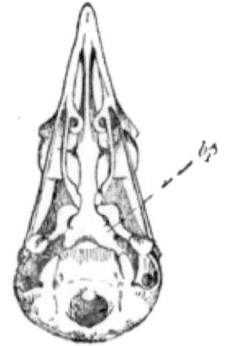

FIG. 1.—Skull of *P. exustus*.   FIG. 2.—Skull of *P. exustus*.

presence of basipterygoid processes (*bp*) in the skull (Fig. 1), the shoulder-girdle, sternum, and especially the great deltoid process of the humerus, or upper-wing bone; but the digestive organs are like those of the True Game-Birds. Among other distinctive characters may be mentioned the "schizorhinal" nasals* and the sternum with *two* notches on each side of the posterior margin, the inner one being sometimes reduced to a foramen (Fig. 4).

The bill resembles that of the True Game-Birds, but is not so strongly developed.

Three toes only occur, the hind-toe, when present, being in a rudimentary condition. The feet are very short and feathered, and the toes are either naked or thickly covered with plumes.

The wings are long and pointed.

The feathers of the body have well-developed after-shafts,

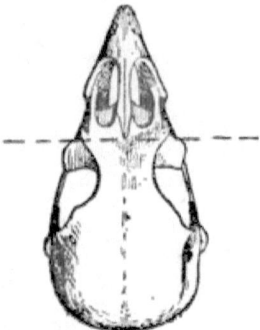

FIG. 3.—Skull of *P. alchatus*.

like those of the True Game-Birds, but the fifth secondary flight-feather is absent.

The young are born covered with down, and are able to run soon after they are hatched.

The eggs are almost invariably three in number, smooth and glossy in texture, equally rounded at both ends, and double

* This character is not of very much importance, as it varies in the different species of Sand-Grouse. In the common Pin-tailed Sand-Grouse (*Pteroclurus exustus*) the backward prolongation of the inter-maxillary bones falls short of the horizontal line drawn between the posterior margins of the nasal notches, so that in this species we see a typical "schizorhinal" skull (Fig. 2). In Pallas' Sand-Grouse (*Syrrhaptes paradoxus*) the inter-maxillary processes reach this line and a perfectly intermediate type ensues; but in the Eastern Pin-tailed Sand-Grouse (*P. alchatus*) they extend beyond, and practically a "holorhinal" form is shown (Fig. 3).

spotted, a set of pale purplish marks beneath the surface of the shell underlying the brown surface spots.

All the Sand-Grouse are birds with immense powers of

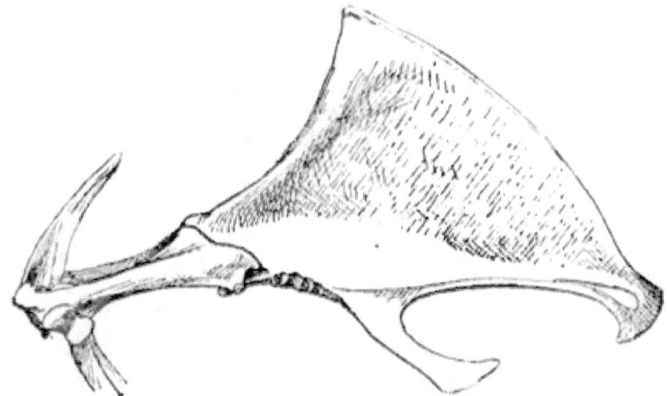

FIG. 4.—Sternum of *P. arenatus*.

flight, able to traverse great distances in a remarkably short space of time.

The majority of the species are migratory, some of them wandering thousands of miles.

THE SAND-GROUSE. FAMILY PTEROCLIDÆ.

THE THREE-TOED SAND-GROUSE. GENUS SYRRHAPTES.

Syrrhaptes, Illiger, Prodr. p. 243 (1811).

Type, *S. paradoxus* (Pall.).

This genus is distinguished from the rest of the Sand-Grouse by the want of the hind-toe. The other toes, as well as the feet, are entirely covered with feathers.

1. PALLAS' THREE-TOED SAND-GROUSE. SYRRHAPTES PARADOXUS.

Tetrao paradoxa, Pall. Reis. Russ. Reichs. ii. p. 712, pl. F. (1773).

Syrrhaptes paradoxus, Ogilvie-Grant, Cat. B. Brit. Mus. xxii. p. 2 (1893).

(*Plate I.*)

Adult Male.—General colour pale sandy. No black spots on the sides of the neck; a band of white feathers across the breast, each feather with a narrow black cross-bar before the tip; throat rust-red, not margined with a black line. *A large black patch on the abdomen.* Total length, 14·6 inches; wing, 9·1; tail, 7; tarsus, 0·8.

Adult Female.—Differs from the male in having the sides of the neck spotted with black; the band across the breast is wanting; a black line bounds the pale buff throat. Total length, 12·8 inches; wing, 8; tail, 5·5; tarsus, 0·8.

Nestling.—Covered with beautifully-patterned down, each plume of the body being distinct and almost scale-like in appearance, quite different from the fluffy down of young Game-Birds. The general colour is pale buff, with patches of sienna and brown arranged in pairs on the sides of the head and upper-parts of the body. These patches are mostly margined and connected by irregular dotted black lines. (Cf. Newton, Ibis, 1890, p. 210, pl. vii.)

Range.—Kirghiz Steppes, throughout Central Asia to Mongolia and Northern China, extending northwards to the north of Lake Baikal and south to Turkestan. A sporadic migrant to Western Europe. Periodically, and from some unknown cause, great numbers visit Europe in early summer, even penetrating to the islands on the western coasts. The first great visitation took place in 1863, and again, in 1888, enormous numbers spread themselves over Europe and bred in various places, both eggs and young having been obtained. In other years smaller flocks have been observed, but none have ever succeeded in establishing themselves permanently, the comparatively small number that escaped being shot or killing themselves on the telegraph wires, having always disappeared, and possibly succeeded in returning to the Kirghiz Steppes.

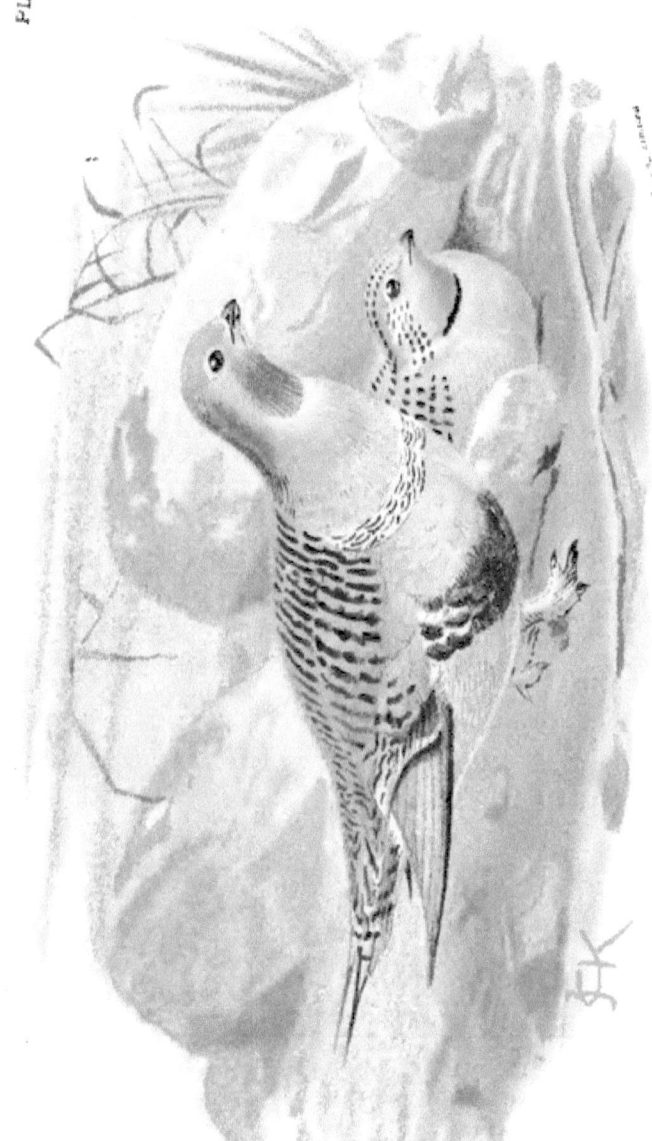

PALLAS' THREE-TOED SAND-GROUSE

Habits.—All the Sand-Grouse are very similar in their habits, drinking in the early morning and evening, and often traversing great distances to reach their accustomed drinking-places. They feed in the morning and afternoon, resting, sunning, and dusting themselves during the heat of the day in fine weather, though on rough stormy days they are generally very unsettled and constantly on the move. Pallas' Three-toed Sand-Grouse does not differ in habits from the rest of its allies, and Prjevalsky gives the following account of its mode of life. " After their morning feed, the flocks betake themselves to some well or salt-lake to drink, apparently preferring the fresh to the salt water. At the drinking-place, as well as at the feeding-places, these birds never settle on the ground without first describing a circle, in order to assure themselves that there is no danger. On alighting they hastily drink and rise again ; and, in cases where the flocks are large, the birds in front get up before those at the back have time to alight. They know their drinking-places very well, and very often go to them from distances of tens of miles, especially in the mornings between nine and ten o'clock, but after twelve at noon they seldom visit these spots." In autumn they are very gregarious and large flocks are to be met with in the neighbourhood of their breeding-ground, unless compelled to migrate to greater distances by a heavy fall of snow.

Swinhoe tells us that in North China great numbers of these birds are sometimes caught after a snow-storm, when they arrive in large flocks in search of food. Having cleared the snow from a patch of ground, the natives scatter a small green bean to attract the birds and sometimes manage to catch a whole flock in their clap-nets.

Nest.—None ; merely a slight hole scratched in the ground.

Eggs.—Like those of all other members of the group, the eggs are perfectly oval in shape and remarkably Rail-like in appearance, closely resembling those of the Corn-Crake (*Crex crex*). The ground-colour is olive or brownish-buff, spotted all over, though not very thickly, with brown and pale lilac or grey, the former markings being on the surface of the shell, the latter beneath. *Three* is the usual number, though it is said that four are occasionally found.

II. TIBETAN THREE-TOED SAND-GROUSE. SYRRHAPTES TIBETANUS.

Syrrhaptes tibetanus, Gould, P. Z. S. 1850, p. 92; Hume and Marshall, Game Birds of India, i. p. 43, pl. (1879); Ogilvie-Grant, Cat. Birds Brit. Mus. xxii. p. 5 (1893).

Adult Male.—Differs from *S. paradoxus* in having the abdomen white, *with no black patch;* the black vermiculations, or narrow wavy markings, on the back of the neck and interscapular region, are very fine, and gradually become obsolete on the wing-coverts and scapulars. Total length, 16 inches; wing, 9·8; tail, 7·9; tarsus, 1.

Adult Female.—Similar to the male, but having the vermiculations and markings equally well-defined on all the upper-parts of the body. Total length, 15 inches; wing, 9·7; tail, 7·4; tarsus, 1.

Nestling.—Closely resembles that of *S. paradoxus*.

Range.—The home of this species is Tibet, where it inhabits the Alpine tracts, from 12,000 to 18,000 feet. In the north it extends to the steppes of Koko-nor, westwards to the Pamir, and southwards to Ladak and the upper portions of the Sutlej Valley. Mandelli obtained specimens from the part of Tibet immediately to the north of Sikhim, but how far east it is met with is somewhat uncertain.

Habits.—The semi-desert plains of the desolate steppes are the favourite haunts of these birds, and there they feed on grass, seeds, and berries, being generally met with in pairs or small flocks. Mr. Hume says: " During the middle of the day it squats about, especially if the day be hot, basking in the sun, very generally scratching for itself a small depression in the soil. Both when feeding and taking its siesta, it is not uncommonly in considerable flocks (I have seen several hundreds together); but in summer, at any rate, it is perhaps more common to meet with it in little parties of from three to twenty. Whilst feeding, it trots about more rapidly and easily than its short feather-encased legs and feet would lead one to suppose, individuals continually flying up and alighting a few yards farther on, and now and again the whole flock rising and flying round, apparently without reason or aim.

"Sometimes it is very shy, especially in the early mornings and evenings; and though it will not, unless repeatedly fired at, fly far, it will not let you approach within one hundred yards; but, as a rule, during the heat of the day, you may walk right in amongst them. They are precisely the colour of the sand when basking, and often the first notice you have of their proximity is the sudden patter of their many wings as they rise and dart away, and the babel of their cries, which, if the flock be a large one, is really startling for a moment. . . . Early in the morning, and quite at dusk, they come down to the water to drink; by preference to fresh water, but, as at the Tso-Khar, at times to quite brackish water.

"They are always noisy birds when moving about, uttering a call something like 'guk-guk,' to my ear, or, again, as some people syllable it, 'yak-yak,' 'caga-caga,' &c., &c., but they are specially noisy in the evenings when they come down to drink."

Nest.—None.

Eggs.—Very similar to those of *S. paradoxus*, but larger.

THE PIN-TAILED FOUR-TOED SAND-GROUSE. GENUS PTEROCLURUS.

Pteroclurus, Bonap. Compt. Rend. xlii. p. 880 (1856).

Type, *P. alchatus* (Linn.).

The five members of this genus are distinguished from *Syrrhaptes* by having the hind-toe always present, though small, and only the feet feathered, *the toes being naked;* from *Ptero-*

cles it differs in having the two middle tail-feathers much longer than the rest and pointed.

1. THE EASTERN PIN-TAILED SAND-GROUSE.   PTEROCLURUS ALCHATUS.

Tetrao alchata, Linn. S. N. i. p. 276 (1766).
Pterocles alchata, Hume and Marshall, Game Birds of India, i. p. 77, pl. (1878).
Pteroclurus alchata, Ogilvie-Grant, Cat. Birds Brit. Mus. xxii. p. 7 (1893).

Adult Male.*—Lower breast and belly pure white, and the

* It is to Mr. E. G. B. Meade-Waldo, who has for many years kept Spanish specimens of the Western Pin-Tailed Sand-Grouse (*P. pyrenaicus*) in captivity, that I owe the following remarkably interesting details regarding the seasonal changes of plumage in this species. His observations are based on a number of specimens kept in his aviary, one male having lived for eight years, and I now find that specimens in the British Museum bear out the correctness of Mr. Meade-Waldo's statements, so far as one is able to judge from specimens which have only been obtained during the winter months of December to February. No doubt, from the absence of material, I misinterpreted the changes of plumage in the fine winter series of birds in the British Museum. Many of the specimens in December plumage, still retaining some barred autumn feathers on the top of the head and back, and with the throat nearly, or entirely, pure white, which I treated in the "Catalogue of the Game-Birds" as *immature* males, I now perceive to be in many cases *adult* males assuming winter plumage. Again, February specimens which I regarded as younger males in a more advanced stage of plumage are mostly adult males moulting into their summer or breeding dress.

Mr. Meade-Waldo writes: "The seasonal changes in the plumage of the *males* of my tame Sand-Grouse (*P. pyrenaicus*) take place in this manner. At the annual complete moult *in the beginning of June*, when all the feathers, as well as those of the wings and tail, are changed, the breeding plumage is exchanged for one of the following pattern : the whole of the back and upper-parts are replaced by feathers of a sandy-yellow with black transverse bars ; the black throat and black stripe behind the eye by white ; the broad chestnut belt on the breast by one much lighter in colour, the black edges of which are indistinct. In this plumage the male closely resembles the female, only wanting the double black bar on the throat and the bluish-grey band on the feathers of the back.

In *September* the back-feathers are replaced by a suit of plain olive without cream-coloured tips.

In *December* the full breeding plumage begins to appear, and is complete by the end of January. The plain olive of the back is replaced by a sub-terminal bars on the wing-coverts white. Throat in *summer plumage* black, in *winter* white, like that of the female; chest in *summer* light rufous, in *winter* much paler and similar to that of the female; upper-parts in *summer* dull olive, with an ochreous patch at the end of most of the feathers; in *autumn* barred with yellowish-buff and black, and somewhat like the plumage of the female at all seasons; in *winter* uniform dull olive, with rarely any ochreous markings. Total length, 14·8 inches; wing, 8·4; tail, 6·3; tarsus, 1·1.

Adult Female.—Distinguished from the male at all seasons by the slate-grey or whitish bands near the extremity of the barred feathers of the back and upper-parts; the throat is white at all seasons; and the chest paler rufous than that of the male in summer plumage, but similar to that of the winter plumage. Total length, 13·5 inches; wing, 7·8; tail, 5·5; tarsus, 1.

Nestling.—Very similar to that of the Pallas' Three-toed Sand-Grouse (*S. paradoxus*), but differing somewhat in pattern and readily distinguished by the naked toes and rudimentary hallux.

Range.—South-western Asia is the home of this species, which is found from Palestine to North-western India, and extends southwards to the head of the Persian Gulf and

paler olive with creamy-yellow tips; the white throat and white stripe behind the eye by black; the barred head by uniform creamy slate-colour, and the chestnut belt on the breast becomes bright in colour, with sharp black bands above and below, caused by the wearing off of the edges of the feathers.

These changes are complete in adult and vigorous male birds, but in some individuals, when the moult happens to be prolonged, those feathers of the back which are *last* cast, and are consequently developed towards the time when the autumn plumage should be assumed, partake of the characters of both the summer and autumn plumages, being olive with more or less distinct yellow and black bars.

The *female* does not appear to go through any of these changes, only changing the white stripe behind the eye for a black one, and a few of the dorsal feathers being replaced about the beginning of January for others with brighter and broader blue-grey bands, while the black bands on the breast and throat become more intense by tip-shedding.

probably Arabia, while to the north it is met with in Asia Minor, Transcaucasia, and Turkestan.

Habits.—In North-western India it is only a cold-weather visitant, arriving in enormous flocks, which are said to far outnumber those of any other species of Sand-Grouse. It does not breed in the Punjab, but takes its departure about the end of March. Severtzov tells us he found it breeding in the Tian Shan and Karatall ranges at elevations of from 1,000 to 4,000 feet, and eggs have been obtained in the neighbourhood of Smyrna. Mr. Hume writes: "I have seen very little of this species myself, and only on a vast plain some miles from Hoti Mardán where, during the winter, they were in tens of thousands. This plain is partly barren, partly fallow, and partly cultivated with wheat, mustard, and the like. It was only on the barren and fallow lands that I saw them. They were extremely wary, and it was only by creeping up a *nala*, or small ravine, that it was possible to get within even a long shot of them. Their flight is extremely rapid and powerful: to me it seemed more so than that of any of their congeners. They are very noisy birds, and whether seated or flying, continually utter their peculiar cry." The food consists of seeds, grain and green leaves, &c.; and though some of the birds shot by Mr. Hume were remarkably fat, when cooked they proved dry and tasteless. Occasionally they are hawked with Shaheens, and sometimes shot by working a couple of Peregrines over them, when they lie very close and are easily approached; but the easiest way of capturing them is in horse-hair nooses.

Nest.—None; merely a depression in the ground.

Eggs.—Very similar to those of *S. paradoxus*; but the ground-colour has a warm reddish-cream tinge, and generally the markings are more profuse.

SUB-SP. *a.* THE WESTERN PIN-TAILED SAND-GROUSE
PTEROCLURUS PYRENAICUS.

Pterocles pyrenaicus, Seebohm, Ibis, 1883, p. 26 (ex Briss.).

Pteroclurus pyrenaicus, Ogilvie-Grant, Cat. Birds Brit. Mus. xxii. p. 9 (1893).

This is merely a darker, and more richly coloured, western form of *P. alchatus.*

Adult Male.—Differs from the male of *P. alchatus* in having the sub-terminal bars to the wing-coverts yellow, instead of white: the chest in summer plumage rich chestnut, instead of rufous.

Adult Female.—Differs from the female of *P. alchatus* in having the sub-terminal bars to the wing-coverts yellow, instead of white; the chest rufous or pale chestnut, very similar to that of the male of *P. alchatus* in summer plumage.

Range.—Southern Europe and North Africa are the home of this form, which is found in suitable parts of Spain and Portugal, as well as the south of France, while it has been obtained at various other localities to the east, as, for instance, Sicily, Malta, Greece, and Cyprus. South of the Mediterranean it is common in Eastern Morocco and the extensive sandy plains at the base of the Atlas Mountains.

Habits.—Similar to those of its eastern ally. Canon Tristram writes: "Though this bird does not approach so near the verge of cultivation northwards as the Black-bellied Sand-Grouse (*P. arenarius*), it is far more generally abundant, and continues to occur in vast flocks in winter in the M'zab and Touarick country."

Eggs.—Rich fawn-colour, covered and sometimes zoned with large maroon-red surface blotches, and pale lilac shell-markings.

II. NAMAQUA PIN-TAILED SAND-GROUSE.   PTEROCLURUS
NAMAQUUS.

Tetrao namaqua, Gmel. S. N. i. p. 754 (1788).
Pterocles namaqua, Bocage, Orn. Angola, p. 396 (1881).
Pteroclurus namaqua, Gurney, ed. Andersson's B. Damaraland, p. 242 (1872), Ogilvie-Grant, Cat. Birds Brit. Mus. p. 10 (1893).

Adult Male.—Lower breast and belly *not* pure white; *the shaft of the first flight-feather white ;* a white and chestnut band separates the vinaceous buff-coloured chest from the brown breast, which shades gradually into buff on the belly. Total length, 12 inches ; wing, 6·6 ; tail, 4·6 ; tarsus, 0·85.

Adult Female.—May be distinguished from the male by having no pectoral band, while the breast and belly are buff, barred with black. Total length, 10·5 inches ; wing, 6·2 ; tail, 4·3 ; tarsus, 0·8.

Range.—South Africa, from the Transvaal in the east to Damaraland and Benguela in the west, and extending southwards to the Great Karroo.

Habits.—Andersson found these birds very abundant in some parts of Damaraland, where immense flocks were observed at the water about eight or nine o'clock in the morning. Before descending to drink, they might be seen circling round the water at a considerable height, and adding to their numbers at almost every turn. He says : " Frequently they make no attempt at a descent until they are directly over the spot they intend to visit, when they suddenly descend with great velocity, at the same time describing more or less of a semicircle before they alight."

III. THE COMMON PIN-TAILED SAND-GROUSE. PTEROCLURUS EXUSTUS.

Pterocles exustus, Temm. Pl. Col. v. pls. 28, 29 [Nos. 354 and 360] (1825); Hume and Marshall, Game Birds of India, i. p. 69, pl. (1878).

Pteroclurus exustus, Ogilvie-Grant, Cat. B. Brit. Mus. xxii. p. 12 (1893).

Adult Male.—The lower breast and belly are *not* pure white, and the shaft of the first flight-feather is *dark*. The chest is uniform vinaceous-buff, divided from the yellowish-buff upper breast by a narrow white and black band ; lower breast and belly deep chestnut-brown. Total length, 13 inches; wing, 7·1; tail, 5·3; tarsus, 0·9.

Adult Female.—Has the chest and upper breast buff, spotted with black; and the belly blackish-brown, closely barred with rufous-buff. Total length, 10 inches; wing, 7·0; tail, 3·6; tarsus, 0·8.

Range.—This bird is found over a wide area, and extends from Senegal in Western Africa through North and East Africa across South-western Asia to the Peninsula of India. To the north, it ranges to Palestine and Central Asia, and south to the Pangani River in East Africa.

Habits.—This is the commonest species of Sand-Grouse, and in India, where in suitable localities it is specially numerous, it is to be found in the sandy districts where vegetation is scarce. From the excellent account given by Mr. A. O. Hume the following lines are borrowed:—"The Common Sand-Grouse, though frequently met with in considerable packs numbering from twenty to two hundred individuals, is never, so far as my experience goes, seen in those enormous flocks which *P. alchatus* and, in a somewhat lesser degree, *P. arenarius*, affect. In all parts of the country where I have shot them, I have most frequently seen them in parties of from five to thirty. . . They live wholly on seeds, and no small seeds seem to come amiss to them. . . From 8 to 10 a.m., according to season, they are off to some stream, river, or tank to drink, and where, or at times when, water is scarce and drinking-places few and far between, very considerable numbers resort to the same place, and afford opportunities for very pretty sport, if several guns lie up at distances of from one to two hundred yards from the pool, and shoot the birds fairly as they come and go, high over head. Their flight is then swift and strong, and they will carry off a great deal of shot. . . Their approach is always notified by their peculiar chuckling, far-reaching double call, which they continually utter during flight. . . After the morning drink, they again resort to their feeding-ground, *not* that where they fed earlier, but much more open and bare ground, ploughed fields, and perfectly open sandy plains." Here, like the rest of their kind, they enjoy a noontide siesta, after which, having again fed for a while, they return to the water for their evening draught. Except when coming and going to their drinking-places, the parties are so scattered that it is not easy to obtain many shots, but they are generally far from shy, and, when squatting, exactly resemble the colour of the ground, whence they rise with great rapidity, and often add considerably to the excitement of a day's sport. In India this species breeds during the greater part of the year, and probably two or more broods are reared in the season.

Eggs.—Very similar to those of *P. alchatus;* but proportionately smaller, and with the ground-colour, which varies from pinkish stone-colour to pale olive-brown or grey, generally paler. Three is the number of eggs generally found in a nest, but sometimes only two are laid. Four and even five have been found in one nest-hole, but they are probably the produce of more than one female. The average measurement is 1·45 inch by 1·03.

IV. SPOTTED PIN-TAILED SAND-GROUSE. PTEROCLURUS SENEGALLUS.

Tetrao senegallus, Linn. Mantissa, p. 526 (1867-71).
Pterocles senegalus, Hume and Marshall, Game Birds of India, i. p. 53, pl. (1878).
Pteroclurus senegallus, Ogilvie-Grant, Cat. B. Brit. Mus. xxii. p. 14 (1893).

Adult Male.—Chest and upper breast uniform pale fawn-colour; the shaft of the first flight-feather *dark*, and the middle of the belly *black*. Neither the upper-parts nor the chest are spotted. Total length, 13 inches; wing, 8; tail, 5·4; tarsus, 1.

Adult Female.—Differs from the male in having the upper-parts of the body and chest ornamented with round greyish-black spots. Total length, 12·5 inches; wing, 7·4; tail, 4; tarsus, 0·9. The only species this bird can be mistaken for is the Coronetted Sand-Grouse (*Pterocles coronatus*) described below, but this can be at once distinguished by its short tail.

Range.—From the Southern Sahara in North Africa through South-western Asia to North-western India.

Eggs.—Similar to those of *P. exustus*, but the markings more sparse.

THE SHORT-TAILED SAND-GROUSE. GENUS PTEROCLES.

Pterocles, Temm. Man. Orn. p. 299 (1815); id. Pig. et. Gall. iii. pp. 238, 712 (1815).

Type, *P. arenarius* (Pall.).

The characteristics of this group are the same as those of *Pteroclurus*, but the middle pair of tail-feathers are not produced, being but little longer than the second pair.

1. THE LARGE OR BLACK-BELLIED SAND-GROUSE. PTEROCLES ARENARIUS.

Tetrao arenarius, Pallas, Nov. Com. Petrop, xix. p. 418, pl. viii. (1775).

Pterocles arenarius, Hume and Marshall, Game Birds of India. i. p. 47, pl. (1878); Ogilvie-Grant, Cat. B. Brit. Mus. xxii. p. 18 (1893).

Adult Male.—Distinguished from all the other species by its large size and *uniform black* belly, *none* of the feathers being edged with white. Throat chestnut, terminated by a black band; the chest and breast uniform dove-grey, and the feathers of the back pale rufous and grey, with rufous-buff or yellow ends. Total length, 13·5 inches; wing, 9·3; tail, 4·2; tarsus, 1.

Adult Female.—Differs from the male in having the throat yellowish white, terminated by a black band; the chest and upper breast pale buff, spotted with black; and the back pale rufous-buff, thickly barred with black. Total length, 13 inches; wing, 8·6; tail, 4; tarsus, 1.

Range.—This species has a wide range, being found in North Africa, Southern Europe, and South-western and Central Asia. In the west it extends to the Canary Islands, in the east to North-west India, while northwards it occurs in the Kirghiz Steppes and Dzungaria, and southwards in the Sahara.

Habits.—This extremely handsome bird is only a cold-weather visitant to India, but during the coldest months of the year, is generally to be met with in enormous flocks where sandy plains stretch far and wide, and water is within easy distance. Their habits generally resemble those of the rest of their kind, and they feed, go to the water, and rest at mid-day, in the same way. In parts of the country, where rivers are far distant, they repair morning and evening to drink at such tanks as have not dried up with the approach of hot weather, and by hiding near such spots, it is not difficult to procure large numbers as they come and go. This species is not known to breed in India, but, to the west, it has been found nesting on the lower plateaux at elevations of from four to seven thousand feet.

Nest.—A slight depression in the soil.

Eggs.—Three in number; light stone-colour or buff, marbled with purple-grey shell-markings, and light brown surface blotches. Size, 1·85 to 2 inches by 1·3 to 1·35.

II. THE BRIDLED SAND-GROUSE. PTEROCLES DECORATUS.

Pterocles decoratus, Cabanis, J. f. O. 1868, p. 413, and 1870, pl. iii.; Ogilvie-Grant, Cat. B. Brit. Mus. xxii. p. 21 (1893).

Adult Male.—Size small. Belly black, most of the feathers *narrowly margined with white.* A broad black bar, edged on both sides with white, passes up the middle of the throat and surrounds the gape; eyebrow-stripe white and black; chest uniform buff-grey. Total length, 8·8 inches; wing, 6·4; tail. 2·7; tarsus, 1.

Adult Female.—Distinguished from the male by having the chin and throat uniform buff, or with a few small black spots; the eyebrow-stripe wanting; and the chest buff, barred with black. Total length, 8·6 inches; wing, 6·1; tail, 2·6; tarsus, 1.

Range.—This species is found over a somewhat restricted area compared with that of the preceding species, and has only been met with about Lake Jipi and Mount Kilimanjaro in East Africa, and westwards on the Wembaere Steppes in Masai-land. It has also been sent from Nassa, in Speke Gulf, on the Victoria Nyanza, by the Rev. E. H. Hubbard.

III. THE VARIEGATED SAND-GROUSE. PTEROCLES VARIEGATUS.

Pterocles variegatus, Smith, Rep. Exped. Centr. Afr. p. 56 (1836); id. Zool. S. Afr. Aves, pl. x. (1838); Ogilvie-Grant, Cat. B. Brit. Mus. xxii. p. 22 (1893).

Adult Male.—Belly *dull rufous;* the feathers of the feet uniform buff, *not* barred with black or brown; under tail-coverts uniform; upper surface of the shaft of the first flight-feather *white ;* upper- and under-parts of the body *spotted with white ;* the eyebrow-stripe, chin, and throat *slate-grey.* Total length, 9·8 inches; wing, 6·3; tail, 2·8; tarsus, 1.

Adult Female.—Distinguished from the male by having the eyebrow-stripe, chin, and throat *pale buff*, and the belly indistinctly *barred with white.* Total length, 9·8 inches; wing, 6·3; tail. ·2·8; tarsus, 0·95.

Range.—South Africa, extending eastwards to the Transvaal and west to Damaraland.

Habits.—Ayres met with numbers of pairs of this beautiful little species along the Crocodile River in June, showing that the breeding-season had commenced. These, like other Sand-Grouse, he tells us, are very tough if cooked fresh, but if kept for nearly a week, become tender and well-flavoured.

IV. THE CORONETTED SAND-GROUSE. PTEROCLES CORONATUS.

Pterocles coronatus, Licht. Verz. Doubl. p. 65 (1823); Gould, Birds Asia, vi. pl. 63 (1851); Hume and Marshall, Game Birds of India. i. p. 57 (1878); Ogilvie-Grant, Cat. B. Brit. Mus. xxii. p. 23 (1893).

Adult Male.—Belly *buff*; feathers of feet uniform, *not* barred with black or brown; under tail-coverts uniform*; upper surface of the shaft of the first flight-feather *white*; the plumage *not* spotted with white; throat yellow, divided for about half its length by a *black bar*, which surrounds the gape, but is interrupted on the middle of the forehead by a whitish patch; the chest and breast *uniform*. Total length, 11 inches; wing, 7·1; tail, 3·2; tarsus, 0·9.

Adult Female.—Distinguished from the male by having the throat yellow *without* a black bar, and the chest and breast *barred with greyish-black*. Total length, 10·3 inches; wing, 6·6; tail, 3; tarsus, 0·9.

The male of this species is similar in general appearance to the male of the Spotted Sand-Grouse (*P. senegallus*) described above, but may be at once distinguished by the short middle tail-feathers, the black on the throat and round the gape, and the absence of black on the belly.

Range.—This species is met with in North-eastern Africa and the south-western portions of Asia, and extends from the Southern Sahara to the extreme north-west of India.

Habits.—Its cry is said to resemble that of the Spotted Sand-Grouse. Canon Tristram met with it in small numbers in the

---

* The female sometimes has a bar or two of black on the under tail-coverts, but always wide apart.

Southern Sahara and found it breeding. Practically nothing is known regarding the habits or precise area of distribution of this bird.

Eggs.—Ashy-white, with a few, almost obliterated, pale brown markings.

V. SMITH'S CHESTNUT-VENTED SAND-GROUSE. PTEROCLES GUTTURALIS.

Pterocles gutturalis, Smith, Rep. Exped. Centr. Afr. p. 56 (1836) and Zool. S. Afr. pl. iii. [male] and pl. xxxi. [female] (1838-9); Ogilvie-Grant, Cat. B. Brit. Mus. xxii. p. 25 (1893).

Adult Male.—Belly uniform deep chestnut; tarsus uniform rufous-buff. *not barred* with black or brown; under tail-coverts uniform dark chestnut; upper surface of the shaft of the first flight-feather dark or dusky, *never* white; eyebrow-stripe and throat pale yellowish-buff; a black band across the neck, and a second one from the gape to the eye. Total length, 12 inches; wing, 8·3; tail, 3·4; tarsus, 1·2.

Adult Female.—Differs in having *no marked* eyebrow-stripe; *no* black band across the throat; that from the gape to the eye brownish; and the lower breast and belly chestnut, barred with black. Total length, 11·6 inches; wing, 8·2; tail, 3·1; tarsus, 1·1.

Range.—Found in South-eastern, Eastern, and North-eastern Africa, from the Transvaal to the highlands of Abyssinia and to the Wembaere Steppes and Masai-land.

Habits.—This large and handsome bird is one of the most plentiful of the Sand Grouse near the Limpopo, and Ayres found it breeding there in June. In the neighbourhood of Potchefstroom, he tells us, they are tolerably plentiful towards the end of winter and beginning of spring, but appear to leave in summer. They are seldom met with singly, generally in companies of from three to a dozen or more, and frequent the bare ground not far from water. Their flight is exceedingly strong, and on the wing they somewhat resemble some of the Pigeons, especially *Columba phæonota*. On the approach of danger, they crouch and lie very close to the ground, being then extremely difficult to see; when disturbed, they do not run, but rise quite suddenly with a loud whirring noise.

Eggs.—Three, placed on the bare ground amongst grass, without the slightest appearance of a nest.

VI. THE MASKED SAND-GROUSE.   PTEROCLES PERSONATUS.

Pterocles personatus, Gould, P. Z. S. 1843, p. 15; id. Voyage of the "Sulphur." Zool. p. 49, pl. 30 (1844); Ogilvie-Grant, Cat. B. Brit. Mus. xxii. p. 26 (1893).

Adult Male.—Belly rufous-buff, closely barred with black; tarsi uniform buff, *not* barred with black or brown; under tail-coverts uniform buff; upper surface of the shaft of the first flight-feather dark or dusky, *never* white; a broad black band surrounding the gape; upper back uniform isabelline-brown. Total length, 11·6 inches; wing, 8·5; tail, 3·7; tarsus, 1.

Adult Female.—Distinguished by having *no* black band round the gape; the *upper back*, as well as the lower breast and belly, barred with black. Total length, 11·5 inches; wing, 8; tail, 3·4; tarsus, 1.

Range.—Peculiar to the island of Madagascar.

VII. CLOSE-BARRED SAND-GROUSE.   PTEROCLES LICHTENSTEINI.

Pterocles lichtensteini, Temm. Pl. Col. v. pls. 25, 26 [Nos. 355, 361] (1825); Hume and Marshall, Game Birds of India, i. p. 65, pl. (1878); Ogilvie-Grant, Cat. B. Brit. Mus. xxii. p. 29 (1893).

Adult Male.—Tarsus *uniform white;* under tail-coverts closely barred with black; a pectoral band of *four* bars, buff, reddish-brown, buff and black; throat spotted with black; chest above the pectoral band narrowly barred with black; the wing-coverts white, narrowly barred with black, and with buff tips. Total length, 10·3 inches; wing, 7; tail, 2·8; tarsus, 1·1.

Adult Female.—Differs from the male and is distinguished from the females of allied forms by having *no pectoral band; the throat thickly spotted with black, to the chin;* the upper breast barred with black; *the tarsus pure white*, and the black bars on the wing-coverts and chest narrow and regular. Total length, 9·7 inches; wing, 7; tail, 2·8; tarsus, 1·1.

Range.—North-eastern Africa and South-western Asia, extending from Kordofan and Nubia to Abyssinia, Somali-land, and the Sük Country. Across Arabia to the western portions of Sind.

Habits.—Like the Painted Sand-Grouse described below, this species is chiefly met with among bush- and thin tree-jungle, and in other respects their habits appear to be very similar.

Eggs.—Heuglin occasionally found "nests" of this species, which, he says, contained "two cylindrical-shaped eggs, much the colour of dirty and faded Peewits' eggs."

VIII. THE DOUBLE-BANDED SAND-GROUSE. PTEROCLES BICINCTUS.

Pterocles bicinctus, Temm. Pig. et Gall. iii. pp. 247, 713 (1815); Ogilvie-Grant, Cat. B. Brit. Mus. xxii. p. 30 (1893).

Adult Male.—Under tail-coverts closely barred with black; a pectoral band of *two* bars, white and black; throat not spotted with black; chest above the pectoral band uniform. Total length, 9·7 inches; wing, 6·9; tail, 3·3; tarsus, 0·9.

Adult Female.—*No pectoral band;* throat spotted with black to the chin, especially on the sides; upper breast and chest rather irregularly barred with black; *tarsus barred with blackish-brown*. Total length, 9·7 inches; wing, 6·6; tail, 3·2; tarsus, 0·9.

Range.—South Africa, extending east through the Transvaal, west to Mossamedes, and south to the Orange River.

Habits.—A common species in many parts of South Africa. Ayres says that "next to *P. gutturalis*, this is the most plentiful Sand-Grouse found near the Limpopo. The greater number of those we saw in June were in flocks, but some few had paired and were breeding." According to Andersson it is the commonest species in Damara and Great Namaqualand, where considerable numbers may be seen during the dry season, at any of the few permanent waters that exist in those countries. Large flocks frequent these pools about dark and during the early part of the night, as well as sometimes at early dawn; they remain only a short time at the water and announce their arrival and departure by incessant sharp cries.

IX. INDIAN PAINTED SAND-GROUSE. PTEROCLES FASCIATUS.

Tringa fasciata, Scop. Del. Flor. et Faun. Insubr. ii. p. 92 (1786).

Pterocles fasciatus, Gray, List B. iii. p. 49 (1844); Hume and Marshall, Game Birds of India, i. p. 59, pl.(1878); Ogilvie-Grant, Cat. B. Brit. Mus. xxii. p. 27 (1893).

Adult Male.—Under tail-coverts closely barred with black; a pectoral band of *three* differently coloured bars, chestnut, white or buff, and black; throat not spotted with black; the chest above the pectoral band uniform yellowish-buff, and *each wing-covert with a white and a grey band near the extremity, sometimes with four alternate white and grey bars.* Total length, 10·8 inches; wing, 6·7; tail, 3·3; tarsus, 0·9.

Adult Female.—*No pectoral band;* only a few black spots *at the base of the throat;* upper breast and wing-coverts with narrow regular bars of black; the feathers of the feet *barred with blackish-brown.* Total length, 10·5 inches; wing, 6·2; tail, 2·9; tarsus, 0·9.

Range—Only found in the Peninsula of India.

Habits.—The habits of this beautiful little Sand-Grouse resemble those of *P. lichtensteini*, and are very different from those of most of the species already mentioned, and, though widely distributed throughout India, Mr. Hume says that it is very local, being chiefly found in the neighbourhood of low rocky bush-clad, or thinly wooded, hills, and in forest-tracts where the ground is stony and broken up by ravines. They seldom stray far from their natural haunts, unless during the dry season, when compelled to do so in search of water. Compared with other Sand-Grouse, they run extremely well, and never associate in huge flocks, seven to ten being the largest numbers flushed at one time. When flushed they seldom fly far, and run for a considerable distance after they have alighted. Excellent sport may be had in localities where they are abundant, for they lie well and are seldom, if ever, wild. Writing from the Central Provinces, Mr. Thompson observes: "I can quite corroborate Dr. Jerdon's observations as to the crepuscular habits of this species. It is quite nocturnal and feeds and goes to water even in the darkest night. I have seen birds arrive at the edge of a plain at dusk, and remain feeding and going to water during the dark hours before the moon got up. I have frequently, too, noted parties of six or seven flitting about noiselessly over an opening in the forest long after sunset.

"During the early part of the rains these birds entirely leave the forests and jungles, and then, all through the rains, live in the open country, exactly as *P. exustus* does, but they are never noisy like the latter.

"Large numbers of Painted Grouse are taken during the rainy season by bird-catchers, who, approaching under cover of a screen made of green leaves and twigs, drop a circular net, suspended to a loop and held out horizontally at the end of a long bamboo, over the birds, which, as a rule, never seem to suspect that there is danger at hand."

Nest.—A slight depression scratched in the soil, sheltered by a tuft of grass or low bush.

Eggs.—Two, but more often three are laid; rarely four. Pale salmon, and sometimes buffy stone-colour, with the usual purple spots and clouds underlying specks and tiny streaks of brownish-red. Measurements average 1·42 by 0·98 inch.

X. THE AFRICAN PAINTED SAND-GROUSE. PTEROCLES QUADRICINCTUS.

Pterocles quadricinctus, Temm. Pig. et Gall. iii. pp. 252, 713 (1815); Ogilvie-Grant, Cat. B. Brit. Mus. xxii. p. 32 (1893).
Œnas bicinctus, Vieillot (*nec* Temm.), Gal. Ois. iii. p. 60, pl. 220 (1825).
Pterocles tricinctus, Swains. B. W. Africa, p. 222, pl. xxiii. [female] (1837).

Adult Male.—Under tail-coverts closely barred with black; a pectoral band of *three* bars, chestnut, white or buff, and black; throat not spotted with black; chest above the pectoral band uniform; and *each wing-covert with one or two separate deep black bars, narrowly edged on each side with white.* The male of this species closely resembles that of the Indian Painted Sand-Grouse (*P. fasciatus*), but may be at once distinguished by the markings on the wing-coverts.

Adult Female.—*No pectoral band;* no spots on the throat; *upper breast uniform buff*, contrasting with the belly, which is barred with white and black; tarsus barred with black.

Range.—Extends from Senegambia in the west, to Abyssinia in the east.

Habits.—Unknown.

THE TRUE GAME-BIRDS. ORDER GALLINÆ.

This Order includes the great bulk of the species commonly known as "Game"-Birds, and may be recognised by the following characters.

The nasals are holorhinal (Fig. 5) and true basipterygoid processes are absent, but represented by sessile facets (*sf*) situated far forward on the sphenoidal rostrum (Fig. 6). The

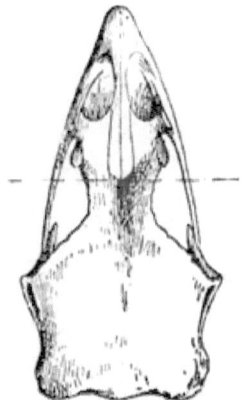
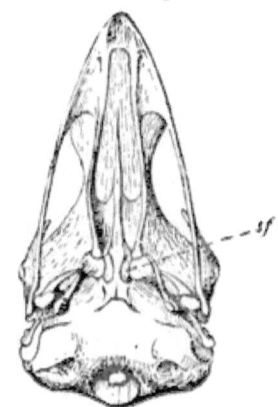

FIG. 5.—Skull of Red Grouse.   FIG. 6.—Skull of Red Grouse.

episternal process of the sternum is perforated to receive a process from the base of the coracoids (Fig. 7, *A*), and there are two deep notches on each side of the posterior margin of the sternum (Fig. 7, *B*).

The bill is short and stout, the upper mandible being arched and overhanging the lower.

The hind-toe is always present, but varies in size and position.

The feathers covering the body are provided with well-developed after-shafts.

The nestlings are born covered with down, and able to run a few hours after being hatched.

The eggs, especially of the smaller species, are often numerous, and when spotted have only a single set of surface marks,

FIG. 7.—Sternum of Red Grouse.

none of the pale underlying spots characteristic of the Sand-Grouse, Hemipodes, and Wading Birds, being found.

THE GROUSE. FAMILY TETRAONIDÆ.

Distinguished by having the hind-toe raised above the level of the other toes. The nostrils are wholly, and the feet (metatarsi) partially or entirely hidden by feathers, never armed with spurs. The toes are either covered with feathers or naked and pectinate, *i.e.*, with a series of horny comb-like processes on each side.

I. THE WILLOW GROUSE AND PTARMIGAN. GENUS LAGOPUS.

Lagopus, Briss. Orn. i. pp. 181, 216 (1760).

Type, *L. lagopus* (Linn.).

These birds may be easily known from all other members

PLATE II

FEATHERS OF SCOTCH GROUSE.

of the *Gallinæ* by having their *feet and toes densely covered with feathers*. The tail is moderately long, and composed of sixteen feathers, the outer ones being nearly as long as the middle pair.

I. THE RED GROUSE. LAGOPUS SCOTICUS.

(*Plates II. and III.*)*

Tetrao scoticus, Lath. Gen. Syn. Suppl. i. p. 290 (1787).
Lagopus scoticus, Leach, Syst. Cat. p. 27 (1816); Millais, Game-Birds, pp. 43-62, pls. and woodcuts (1892); Ogilvie-Grant, Cat. B. Brit. Mus. xxii. p. 35 (1893); id. Ann. Scot. Nat. Hist. 1894, pp. 129-140, pls. v. vi.

Adult Male and Female.—This species may be distinguished by having the flight-feathers *always blackish-brown*.

Male: Total length, 15.5 inches; wing, 8.1; tail, 4.8; tarsus, 1.4.

Female: Total length, 15 inches; wing, 7.8; tail, 4.3; tarsus, 1.35.

Range.—Great Britain and Ireland. The only species of Game-Bird peculiar to the British Islands.

Changes of Plumage.—As no group of birds, as far as we are aware, go through so many and such varied annual changes of plumage as the members of the genus *Lagopus*, which includes the Red Grouse, Willow Grouse, and four species of Ptarmigan, it will be necessary to enter somewhat fully into details so as to thoroughly understand the subject.

The Red Grouse being one of the most variable birds in existence, we must begin by saying a few words regarding individual variation. The ordinary varieties of the *male* may be divided into three distinct types of plumage: a *red form*, a

---

* I am much indebted to the courtesy of the editors of the "Annals of Scottish Natural History" for allowing me to reproduce the plates illustrating my article "On the Changes of Plumage in the Red Grouse," published in their magazine and quoted above.

black form, and a *white-spotted form*. The first of these, in which the general colour is rufous-chestnut (Pl. II., Fig. 8) without any white spots on the breast, is mostly to be found on the low grounds of Ireland, the west coast of Scotland, and the Outer Hebrides. Typical examples of the second, or black form (Pl. II., Fig. 10) are rarely met with, and are usually found mixed with either the red or white-spotted forms, but most often with both, and specimens in mixed plumage are those most commonly met with. The third, or white-spotted form, has the feathers of the breast and belly, and sometimes those of the head and upper-parts, tipped with white. The most typical examples of this variety are found, as a rule, on the high grounds of the north of Scotland.

In the *female*, no less than *five* distinct types are recognisable, the *red*, the *black*, the *white-spotted*, the *buff-spotted*, and the *buff-barred*, forms. The first two are the rarest, the latter being extremely uncommon (Pl. III., Figs. 5 and 13). The white-spotted form occurs as in the male; the buff-spotted form, which is much the commonest and most usually met with, has the feathers of the upper-parts spotted at the tip with whitish-buff (Pl. III., Figs. 2 and 3); the fifth, or buff-barred form (Pl. III., Fig. 4), is met with in the south of Ireland, and resembles in winter (autumn plumage) the ordinary female in breeding plumage, having the upper-parts coarsely barred with buff and black. Very little is known of this last variety, owing to the difficulty in obtaining birds, except during the shooting-season.

The great peculiarity of the Red Grouse, and one without parallel among birds even of the genus, lies in the fact that the changes of plumage in the male and female occur at different seasons.

The *male* has no distinct summer plumage, but has distinct autumn and winter plumages, and retains the latter throughout the breeding-season.

The *female* has a distinct summer plumage, which is complete by the end of April or beginning of May; also a distinct autumn plumage, which is retained till the following spring.

To put it more concisely, both male and female have two distinct moults during the year, but in the male they occur in autumn and winter, and in the female in summer and autumn ; the former having no distinct summer, and the latter no distinct winter, plumage.

In the Willow Grouse and Ptarmigan there are *three* distinct changes of plumage in summer, autumn, and winter in both male and female alike, the winter plumage being *white* in all.

The Red Grouse is considered by most ornithologists merely an insular form of the Willow Grouse, and consequently one might naturally suppose that as the British species does not turn white in winter, such protective plumage being unnecessary in the localities it inhabits, the winter moult has been gradually dropped. Now this is the case with the female only, and we find the male, for no apparent reason, changing his newly acquired buff and black autumn plumage for a winter one of chestnut and black. Further investigations may lead to some explanation of this strange anomaly, but at present we know of none.

Adult Male.—Autumn Plumage.—After the breeding-season a very complete autumn moult takes place, the quills, tail, and feathers on the feet being entirely renewed. In most examples the feathers of the upper-parts are black, margined and irregularly barred with tawny-buff, and in most cases the bars cross the feathers more or less transversely (Pl. II., Fig. 4), but in some they are more or less concentric and parallel with the marginal band, giving the upper-parts a scaled appearance. (Pl. II., Figs. 6 and 7.) The feathers of the chest are rather widely barred with buff or rufous-buff and black (Pl. II., Fig. 11), and some of the flank-feathers are more narrowly barred with the same colours. The rest of the under-parts vary according to the type to which the individual belongs, being chestnut, black, or white-spotted, or a mixture of all three. In a bird shot on the 6th of June, the autumn moult having commenced on the upper mantle, three different sets of feathers can be seen on the back at once, belonging to the new autumn, the old winter, and the old autumn plumages, both the latter very clearly showing the result of wear and tear (Pl. II., Figs. 1-3).

The males at this season, no matter to what type they belong, bear a much closer resemblance to one another than they do in their winter plumage, only the under-parts of the body differing conspicuously.

The first feathers of the winter plumage begin to appear about the beginning of September.

Adult Male.—Winter-Summer Plumage.—General colour above black, with finely mottled bars of dark chestnut (Pl. II., Fig. 5); head, neck, and chest (Pl. II., Fig. 12) mostly dark chestnut, finely marked with black; and the flanks mottled and barred with the same colours, the chestnut usually predominating. Generally a greater or less number of autumn feathers are retained, and are conspicuous among the new winter plumage. The rest of the under-parts remain the same as after the autumn moult.

The general colour of each bird varies, of course, according to the type to which it belongs, some being darker, some lighter. When once the winter moult is complete, *no change whatever* takes place in the plumage of the male till the following autumn moult, except that the feathers become bleached and worn at the extremities.

Adult Female.—Autumn-Winter Plumage.[*]—Upper-parts black,

[*] The form described is the commonest or *buff-spotted* form of the female in autumn plumage. In typical examples of the red form the buff spots at the ends of the feathers of the upper-parts are absent, and this is also the case in the much rarer black form. In the buff-barred form, from the south and west of Ireland, the terminal buff spot takes the form of a marginal bar, and the feathers are practically indistinguishable from the breeding or summer plumage. It may transpire that, in the south of Ireland, the most southerly point of this bird's range, the female retains her breeding plumage throughout the year, but this seems unlikely, and birds killed between the months of April and August are wanted to settle this point.

PLATE III.

with narrow irregular bars and mottlings of rufous, and a buff spot at the tip of most of the feathers (Pl. III., Figs. 2 and 3); chest and flank-feathers narrowly and often irregularly barred with rufous and black, and usually more or less tipped with buff (Pl. III., Figs. 10 and 11). The rest of the under-parts are dark chestnut, mottled and barred with black, or black, barred with chestnut. The typical white-spotted form differs, of course, in having the feathers of the under-parts widely tipped with white.

Adult Female.—Summer Plumage.

A. Feathers of the Upper-parts.

So far as I have been able to ascertain from examining a large number of specimens, the summer feathers of the *upper-parts* are always attained by moult, and never by change of pattern. The summer moult of these parts is very complete, and the transformation from the autumn-winter plumage very remarkable. Every female assumes the summer plumage, and at this season all the different types closely resemble one another, but one can generally tell by the colour of the under-parts to which form an individual belongs. In the average female in full breeding dress the upper-parts may be described as black, each feather being rather widely margined, barred, and marked with orange-buff (Pl. III., Fig. 1). The protection afforded by this plumage is so perfect that, when the bird is sitting on its nest among heather and dead grass, it may easily remain unobserved, though only a few yards distant.

This plumage, however, varies much in different individuals, birds from the west of Scotland, Yorkshire, and Ireland having the orange-brown bars much brighter and wider than in the more finely mottled and darker specimens generally characteristic of the east of Scotland.

B. Feathers of the Sides and Flanks.

By the first week in May the summer plumage of the female

Grouse is fairly complete, and many of the finely mottled rufous and black autumn flank-feathers are replaced by widely, and often irregularly, barred buff and black feathers, similar to those of the chest. It must be particularly noted that in *none* of the many females examined, in breeding plumage, were the *whole* of the autumn flank-feathers cast or changed in the summer moult, a large proportion being retained, unchanged in colour, till the next (autumn) moult. The summer flank-feathers are produced in two ways, either by a gradual re-arrangement and change in the pigment of the autumn feathers (Pl. III., Figs. 6-8) or by moult (Pl. III., Fig. 9). In some birds the whole of the alteration in the plumage of the flanks is produced by change of pattern in the old autumn feathers, in others the change is entirely produced by moult, while sometimes both methods are employed by the same individual. In the former case, the first indication of the coming change may be observed in the beginning of November, or even earlier, when many of the flank-feathers show traces of an irregular buff stripe or spot near the terminal half of the shaft (Fig. 7). As the bird only changes about half its flank-feathers, these buff marks are only to be observed on such as are destined to undergo alteration of pattern, which, roughly speaking, means every second or third feather. The buff spot gradually enlarges and spreads along the shaft, then becomes constricted at intervals and broken up into patches which gradually extend laterally towards the margins of the webs, forming wide irregular buff bands (Fig. 8). Meanwhile the interspaces become black, and the rufous of autumn dies out.

When the summer feathers are supplied by moult, they usually begin to make their appearance about the beginning of March, and even when fully grown, may generally be recognised from those produced by change of pattern, by their more regular black and buff barring (Pl. III., Fig. 9) The change of pattern without a moult appears to take a long time to become complete, for we find, as already shown, that though autumn feathers, altered in this way, begin to show traces of the coming metamorphosis as early as the beginning of November, the colours are often imperfectly arranged by the end of April. When the summer feathers are supplied entirely by moult, no change whatever is visible in the autumn plumage of the flank-feathers till about the end of February, when the first new feathers begin to appear, though we have noted a single instance of one summer feather making its appearance as early as the middle of December.

There can be no doubt that the male completes his autumn moult very much more quickly than the female does, many males being in full autumn plumage by the beginning of September. Possibly this may be accounted for by the resources of the female being more severely taxed than those of the male during the breeding-season. It may very naturally be asked why some females should change their summer flank-feathers by moult, while others are enabled to arrive at the same result by going through the much less exhaustive process of redecorating their old autumn feathers, and making them serve the purpose of new breeding plumage. This is a difficult question to answer, but it seems natural to suppose that the more vigorous birds gain their summer flank-feathers by moult, while nature has enabled the weaker individuals to obtain the necessary protective nesting plumage by a more gradual and less exhaustive process.

C. Feathers of the Chest.

The summer change of the feathers of the fore-neck and chest in the female Red Grouse is similar to that which takes place on the sides and flanks, but is very much more complete, *all* the feathers being widely barred with black and yellowish-buff by the beginning of May (Pl. III., Fig. 12).

As will be easily understood, these being conspicuous parts of the bird when she is sitting on her eggs, it is most important for her that the protective black and buff plumage should be complete. The greater part of this change is generally produced by moult; but, as is the case with the flank-feathers, some individuals (probably less robust females) attain the change without moulting. The same rearrangement of the pigment described in speaking of the flanks takes place in the chest-feathers, and the finely mottled and barred rufous-and-black autumn plumage becomes widely barred with black and buff.

Young birds in July resemble the adult female in breeding plumage in their general colour, but the flank-feathers of the adult plumage begin to appear about this time. By the month of November the young are generally not to be distinguished from the adults.

Nestling.—In this and all the other species of *Lagopus*, the nestling is covered with fluffy yellow down, with rich brown pattern on the upper-parts.

Habits.—This species inhabits the open moors covered with heath and ling from sea-level, but is not found above the limits where these plants grow, its place being taken on the mountain tops of many parts of Scotland by the Ptarmigan. Unlike the Black Game, the Red Grouse is strictly monogamous, each male pairing with one female only, and assisting her to rear the young. The nesting-season is, roughly speaking, April and May, but varies according to locality and season, eggs being sometimes found much earlier and as late as June, though the latter are probably second sittings, the first having been destroyed. The female in her black-and-buff summer garb is practically invisible when sitting on her nest, her colours harmonising perfectly with her surroundings.

As the young Grouse becomes strong on the wing and the season advances, the various coveys, especially if the weather is wet and stormy, soon unite their forces and go about in large flocks known as "packs," the males and females generally forming separate parties; and it is not uncommon to find that all the birds killed in one drive are cocks, while on another beat the reverse obtains.

Grouse-shooting commences on the 12th of August and ends on the 10th of December. During this period enormous numbers of birds are shot, the great majority by driving. In Yorkshire and other parts of the north of England where the moors are of large extent and comparatively level, the birds pack so early in the season, and are then so wild, that driving them is the only means of obtaining a bag. From a sporting point of view, it is hardly necessary to add that the superiority of birds driven at a headlong pace over the guns, as compared with those walked up and shot as they rise, is beyond all question. On some of the rougher moors, when driving is impossible or nearly so, one may still have the pleasure of seeing dogs used to find the birds, but unfortunately this form of sport is rapidly going out of fashion. In the west of Ross-shire, the Isle of Skye, and the Hebrides the tameness of the Grouse is well-known, and in fine weather the birds lie as close in December as at the beginning of the season, remaining in small coveys and often sitting till nearly trodden on. Grouse are extremely fond of grain, and during the autumn may generally be seen in the morning and evening in numbers on stubble-fields within reach of the moors they inhabit. Periodically the moors are devastated by a terrible scourge known as "Grouse-disease," which sometimes destroys the greater part of the stock in the localities affected. It is now generally agreed that over-stocking is the primary cause, and the disease is almost always most severe in the springs which follow unusually good seasons, when birds have been particularly numerous and were not sufficiently killed down. The liver and intestines are the parts attacked, the former becoming like dull red jelly and of about the same consistency. Although parasitic worms are usually specially numerous in birds which have died of the disease, they are in no way the cause of death and are often numerous in perfectly healthy individuals.

The Red Grouse occasionally interbreeds with the Black-cock (*Lyrurus tetrix*) and perhaps with the Ptarmigan (*L. mutus*) but the supposed hybrids with the latter species are possibly merely partial albinos of the Red Grouse. Mr. J. G. Millais records and figures a singular hybrid between this species and a Bantam Fowl !

Nest.—A slight hollow in the ground, sheltered by the longer heather and grass, and lined with moss and grass or such materials as chance to be on the spot.

Eggs.—Vary in number from seven to ten and sometimes more. The ground colour is pale cream or buff, spotted and blotched all over with dark reddish-brown, which often nearly conceals the ground-colour. Average measurements, 1·75 by 1·32 inches.

II. WILLOW GROUSE, OR RIPA.   LAGOPUS LAGOPUS.

Tetrao lagopus, Linn. S. N. i. p. 274 (1766).
Tetrao albus, Gmel. S. N. i. pt. ii. p. 750 (1788).
Tetrao saliceti, Temm. Pig. et Gall. iii. pp. 208, 709 (1815) [part].
Lagopus albus, Steph. in Shaw's Gen. Zool. xi. p. 292 (1819); Dresser, B. Europe, v. p. 183, pls. 483, 484 (1874).
Lagopus lagopus, Bendire, Life Hist. N. Am. B. p. 69, pl. ii. figs. 5-10 (1892); Ogilvie-Grant, Cat. B. Brit. Mus. xxii. p. 40 (1893).

Adult Male and Female.—Outer tail-feathers black, with only the bases and tips more or less white ; the flight-feathers *always white ;* the *bill much larger and stouter, like that of L. scoticus*, and the wing about 8 inches in length from the bend to the tip of the flight-feather.

Male : Total length, 15·5 inches ; wing, 8·1 ; tail, 4·8 ; tarsus, 1·4.

Female : Total length, 15 inches; wing, 7·8 ; tail 4·3 ; tarsus, 1·35.

Range.—Circum-polar, inhabiting the Arctic tundras of Europe, Asia, and America.

Adult Male and Female.—Winter Plumage.—Pure white, with the exception of the black outer tail-feathers, which remain unchanged.

Adult Male.—Summer Plumage.—The head and neck chestnut, shading into dark chestnut, or sometimes even black on the chest; rest of the upper-parts chestnut, mottled and barred with black, and often tipped with buff; flight-feathers and rest of under-parts white, as in winter. This is the most complete form of summer plumage found in birds inhabiting the more temperate parts of the range; in those from high altitudes, all the upper-parts, from the back of the neck, remain white, merely interspersed here and there with a few feathers of the summer plumage.

Adult Female.—Summer Plumage.—Very similar to the female of the Red Grouse in breeding plumage, but the buff markings are paler and more conspicuous and the flight-feathers are *white*. Unlike the male, the summer moult of the female, no matter the locality, is always complete, birds from the far north of Alaska being in quite as complete breeding-dress as those from more southern latitudes.

Adult Male and Female.—Autumn Plumage.—Head, throat, and chest light brick or pale chestnut colour, usually with finely mottled black cross-bars (in the female these parts are generally largely intermixed with the old summer feathers); the upper-parts are black, with narrow bars of rufous or rufous-buff. The flight-feathers, tail-feathers, and feathers of the feet are, as in other members of this genus, renewed at this season.

Quite young birds have the first flight-feathers greyish-brown, mottled with buff at the tip and along the outer web.

Habits.—These birds in every way resemble the Red Grouse. Their call is the same, and their eggs are indistinguishable, but they inhabit somewhat different ground, being chiefly found among birch- and willow-trees, and, unlike the Grouse, they are fond of perching on trees, and prefer to roost in them.

NOTE.—The Newfoundland bird is said to differ in having black shafts to the flight-feathers, and has been distinguished under the name of *Lagopus alleni*, but specimens recently obtained from that island show that this difference is not constant, and of no importance. Black shafts to the primary flight-feathers, usually accompanied by black on the adjacent parts of the web, are characteristic of younger birds in the first white winter plumage, but are sometimes to be seen in individuals which are certainly more than one year old.

Nestling, Nest, and Eggs.—Like those of *L. scoticus*.

III. THE COMMON PTARMIGAN. LAGOPUS MUTUS.

Tetrao lagopus, Scop. (*nec* Linn.) Ann. i. p. 118 (1769).
Tetrao mutus, Montin, Phys. Sälsk. Hand. i. p. 155 (1776-86).
Lagopus mutus, Leach, Syst. Cat. p. 27 (1816); Millais, Game
    Birds, pp. 63-72, pls. and woodcuts (1892); Ogilvie-Grant,
    Cat. B. Brit. Mus. xxii. p. 45 (1893).

Adult Male and Female at all seasons.—Outer tail-feathers black, with only the bases and tips more or less white; flight-feathers *always white; bill much more slender* than in the Red Grouse or Willow Grouse; wing shorter, males measuring about 7·5 inches from the bend of the wing to the end of the longest flight-feather.

Adult Male and Female.—Winter Plumage.—General plumage and middle pair of tail-feathers white, with a black patch in front of the eye in the *male*, which is absent or rudimentary in the *female*.

Adult Male.—Summer Plumage.—Head, upper-parts, middle pair of tail-feathers, sides, and flanks dark brown, mottled and barred with grey and rusty; breast brownish-black, sometimes more or less barred and mottled with buff; rest of under-parts white.

Adult Female.—Summer Plumage.—General colour above black, mixed with rufous-buff, most of the feathers being edged with whitish-buff; middle pair of tail-feathers and under-parts rufous-buff, barred with black.

Adult Male and Female.—Autumn Plumage.—Upper-parts, middle pair of tail-feathers, breast, and sides grey, finely mottled with black, and sometimes with buff; rest of under-parts white. The *female* may generally be distinguished by having some feathers of the faded summer plumage remaining among the grey autumn plumage.

Male: Total length, 14·5 inches; wing, 7·6; tail, 4·6; tarsus, 1·3.

Female: Total length, 14 inches; wing, 7·4; tail, 4·1; tarsus, 1·3.

Range.—The mountains of Europe, and possibly also some of the ranges of Central Asia, are the home of the Ptarmigan, but the birds found in the latter localities should, perhaps, be referred to the more northern rufous form, *L. rupestris*, which was the bird found by Mr. Seebohm on the Yenesei at $71\frac{1}{2}°$ N. latitude. In the west it ranges to the mountains of Scotland, in the south to the Pyrenees and Alps, and in the east at least as far as the Ural Mountains.

Changes of Plumage.—Mr. J. G. Millais, who has had exceptional opportunities of studying the plumage of the Ptarmigan from different parts of Scotland, gives the following excellent account of the various changes during the year:—

"*January.*—The white plumage.

"*February.*—The same. (In very early spring the first summer-plumage feathers begin to appear, always on the neck.)

"*March and April.*—Summer plumage coming gradually in, the breast-feathers being the last to appear.

"*May.*—Summer plumage quite complete by the last week of the month.

"*June.*—Summer plumage. Males generally showing white tips to feathers.

"*July.*—The white tips to the feathers of the back and breast in the male have now worn off the feathers, the breast being very black and the whole plumage much darker, and in the female the whole plumage is more rusty and faded. During the last week of the month many of the blue-grey feathers of the autumn make their appearance, and the feathers moult from off the legs. This is the case with both cock and hen, but some specimens are far more advanced than others.

"*August.*—A complete change of both cock and hen to blue-grey plumage of the autumn, the whole being complete about the 20th of this month. The hens sometimes retain a few of the faded summer-plumage feathers till the first week in September, most noticeably on the back and flanks. At the beginning of this month the head and neck are more or less dark, with a brown tinge in both sexes, but by the end of the month the whole bird has changed to a very much paler blue-grey, the black ribbings on the feathers becoming less distinct. At the end of the month the feet are covered with the new feathers, though these are short.

"*September.*—The fading of the feathers in both male and female continues throughout the month; the males exhibit a slight difference in the ground-colouring of the back-feathers, some retaining the brownish tinge and others a pure blue-grey. In the brown-tinged birds the black markings on the feathers are always far less distinct than in the grey birds.

"*October.*—The plumage of both sexes still continues to fade, while the black markings become less and less distinct, till the middle of the month, when the first pure white feathers of the third moult make their appearance. These first show on the back and flanks of the birds, and gradually increase, till by the end of the month both male and female have an equal proportion of both old and new feathers. The feet are by this time quite fully covered, the feathers having been gradually growing since the beginning of August.

"*November.*—A few of the old feathers of the autumn remain on the back, and one or two on the head. Those on the back have, by the 15th of this month, become so pale that the small black markings across them can hardly be discerned, but those on the head and neck do not fade much. At this season, most of the birds will have cast the last of these old feathers, and will stand complete in their new winter dress, in which they continue until the end of February.

"*December.*—The different plumages noted in the preceding month may be more or less normal, but during this month many birds, especially the cocks, retain throughout the winter a large amount of the autumn feathers on the back. One that I received on December 31st, 1890, from West Ross-shire, is figured (*l.c.*) as an example of this stage, and may be taken as a typical specimen, though somewhat dark. In December the average of pure white birds is about one in four, but in severe winters they vary materially, and all the birds may be pure white."—*Game-Birds and Shooting-Sketches*, pp. 69, 70 (1892).

Habits.—The home of the Ptarmigan is among the high stony table-lands and rocks above the limits of tree-growth and heaths. Like the Willow Grouse, the plumage of the male varies greatly in different localities, and the amount of white feathers retained during the summer and autumn plumages is greatly affected by the latitude which the birds inhabit, examples from the north of Norway retaining much white in the upper-parts throughout the summer months. This does not apply to the females, all of which get their full summer breeding-dress, which is no doubt essential for their protection during the nesting-season. In the same way, the mixed plumage of the males no doubt renders them less conspicuous among the patches of snow which, in the more northern latitudes, are not melted during the short summer. The general habits of the Ptarmigan resemble those of the Grouse, their monogamous habits, mode of nesting and feeding, being much the same, but the call is very different from the "bec" of the latter, and is more of a hoarse croak. The female is an excellent mother, taking the greatest care of her young, and boldly menacing any unexpected intruder who may come on her unawares. She flutters along the ground or runs towards her supposed enemy with drooping wings and halting gait to attract attention, while the young disappear as by magic, and vanish among the crevices of the stones. Ptarmigan depend greatly for safety on the perfect harmony of their plumage with their natural surroundings, and it is astonishing to see how they will sometimes rise all round one, almost from under one's feet, on comparatively bare ground, without any previous evidence of their presence.

Nestling, Nest, and Eggs.—Similar to those of the Red Grouse, but the eggs of the latter are rather smaller, less thickly covered with blotches, and more buff in general appearance.

IV. THE ROCK PTARMIGAN. LAGOPUS RUPESTRIS.

Tetrao rupestris, Gmel. S. N. i. pt. ii. p. 751 (1788).
Lagopus rupestris, Leach, Zool. Misc. ii. p. 290 (1817); Bendire, Life Hist. N. Am. B. p. 75, pl. ii. figs. 11-15 (1892); Ogilvie-Grant, Cat. B. Brit. Mus. xxii. p. 48 (1893).
Lagopus rupestris reinhardti, p. 78; *L. r. nelsoni*, p. 80; *L. r. atkhensis*, p. 81; *L. welchi*, p. 82; Bendire, Life Hist. N. Am. B. (1892).

(*Plate IV.*)

It appears to me more and more doubtful whether this so-called species should be considered more than a mere climatic variety of the Ptarmigan. In typical examples, the summer and autumn plumages are certainly more rufous in birds from Iceland, Greenland, Arctic America, Japan, and Asia; but in Newfoundland we find a greyer form, apparently scarcely to be distinguished from the European bird, and similar forms are recorded from some of the islands to the north of Arctic America. Insufficient material prevents us at present from settling this point, but we believe that the most reasonable way of treating the matter is to regard all as climatic variations of one circum-polar species. It is obviously useless to give endless names to slight climatic varieties because they occur in different parts of the globe, when,

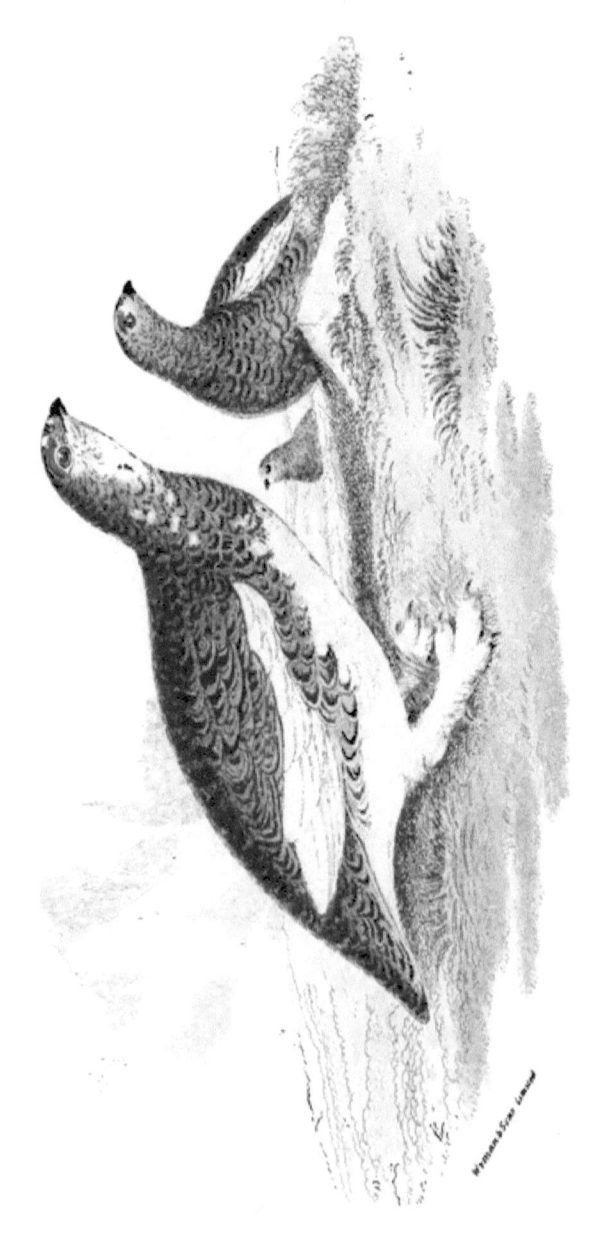

unless one knows the locality whence each individual is obtained, it is practically impossible to name a specimen. Birds from Iceland, Arctic America, Japan, and North Asia are indistinguishable in summer and autumn, and those from Europe, Scotland, and, apparently, also from Newfoundland and some of the islands to the north of Arctic America are equally so, while examples from Greenland have the markings somewhat finer than in North American birds, though, like them, they belong to the more rufous form. In the white winter plumage, all the forms are, of course, perfectly similar to one another. The reader must judge for himself which view of the matter is the most natural. He can regard the Ptarmigan and Rock Ptarmigan as forming one widely distributed species with various climatic phases of grey or rufous plumage, which occur in scattered localities; or he may consider each local form as representing an incipient sub-species or race, but, from all we at present know, the former view seems preferable. The mere fact that indistinguishable grey or rufous forms are found in intermediate localities over a very wide range, seems to show that only one polymorphic species really exists. Among many parallel instances we may mention the little Hemipode (*Turnix taigoor*) found in India and the Indo-Chinese countries.

V. THE SPITSBERGEN PTARMIGAN. LAGOPUS HYPERBOREUS.

Lagopus alpina, var. *hyperborea*, Sundev. in Gaim. Voy. Scandin. Atl. livr. xxxviii. pl. (1838).

Lagopus hemileucurus, Gould, P. Z. S. 1858, p. 354; Dresser, B. Europe, v. p. 179, pl. 482 (1871).

Lagopus hyperboreus, Ogilvie-Grant, Cat. B. Brit. Mus. xxii. p. 51 (1893).

Adult Male and Female.—This species may be easily distinguished, at all seasons of the year, from the other Ptarmigan, by having much more white on the basal part of the tail-feathers. The second pair has the basal two-thirds of both webs white, and the outermost pair shows at least the basal two-thirds of the outer web white; on the median tail-feathers, the amount of white decreases, being confined to the basal third of the outer web in the seventh pair.

Range.—Only known to occur in Spitsbergen.

Habits.—According to Mr. Abel Chapman, the cry of this bird differs from that of the Common Ptarmigan and resembles the "bec" of the Red and Willow Grouse, instead of the hoarse croak of the Ptarmigan.

VI. THE WHITE-TAILED PTARMIGAN. LAGOPUS LEUCURUS.

Tetrao (Lagopus) leucurus, Swains. and Richards. Faun. Bor.-Amer. ii. p. 356, pl. 63 (1831).

Lagopus leucurus, Bendire, Life Hist. N. Am. B. p. 83, pl. ii. figs. 16, 17 (1892); Ogilvie-Grant, Cat. B. Brit. Mus. xxii. p. 52 (1893).

Adult Male and Female.—Distinguished from the allied species *at all seasons* by having the outer tail-feathers *pure white*, as well as by its smaller size.

Male: Total length, 12·6 inches; wing, 7·3; tail, 4·3; tarsus, 1·2.

Female: Total length, 12 inches; wing, 7·2; tail, 3·7; tsus, 1·4.

Range.—Only met with towards the summits of the Rocky Mountains, from Alaska southwards to the north of New Mexico.

Changes of Plumage.—Very similar to those of *L. mutus* and *L. rupestris*, but the black markings on the *summer* plumage of the *male* are much bolder, and in winter the black mark in front of the eye is *absent*.

Habits.—The White-tailed Ptarmigan, Capt. Bendire tells us, is "a resident and breeds wherever found, rarely leaving the mountain summits, even during the severest winter weather, and then only descending 2,000 or 3,000 feet at most, seldom being found at a lower altitude than 8,000 to 9,000 feet at any time." In the Rocky Mountain region it is generally known by the very appropriate name of the "White" or "Snow" Quail. Grinnell writes: "On the high plateaux where this bird is found, the wind often blows with a tremendous sweep and is almost strong enough to throw down a man. When such a wind is blowing, the Ptarmigan dig out for themselves little nests or hollows in the snow-banks, in which they lie with their heads toward the wind and quite protected from it." In general habits this species is very similar to the Common Ptarmigan, but apparently it is mostly found in small parties of about a dozen, and even in late autumn is rarely met with in packs.

Eggs.—Creamy-buff to pale reddish or salmon-buff; the markings, generally small and well-defined, varying in colour from reddish-brown to chocolate-brown. "They resemble far more the eggs of *Dendragapus* than *Lagopus*" (*Bendire*). Average measurements, 1·75 by 1·2 inches.

THE BLACK GROUSE. GENUS LYRURUS.

Lyrurus, Swains. Faun. Bor.-Amer. ii. p. 497 (1831).

Type, *L. tetrix* (Linn.).

Characterised by having the feet feathered, but, unlike *Lagopus*, the toes are naked and pectinate on the sides. The tail is composed of eighteen feathers, and in the male the outer pairs, which are much the longest, are curved outwards at the extremity.

Only two species are known.

1. THE BLACK GROUSE. LYRURUS TETRIX.

Tetrao tetrix, Linn. S. N. i. p. 274 (1766); Millais, Game-Birds, pp. 21-42, pls. and woodcuts (1892).

Lyrurus tetrix, Swains. and Richards. Faun. Bor.-Amer. ii. p. 497 (1831); Ogilvie-Grant, Cat. B. Brit. Mus. xxii. p. 55 (1893).

Adult Male.—Plumage mostly black; the *under tail-coverts pure white*. Total length, 23.5 inches; wing, 10.3; tail, 8.8; tarsus, 1.9.

Adult Female.—Plumage mostly rufous and buff, barred with black, the black bars on the breast being much coarser than in the female of *L. mlokosiewiczi*, and the tail shorter. Total length, 17 inches; wing, 8.9; tail, 4.5; tarsus, 1.6.

Nestling.—Covered with yellowish down, patterned with chestnut-brown on the upper-parts.

Range.—The common Black Grouse is found in suitable localities over the greater part of Europe and Northern and Central Asia. To the west it extends to Great Britain, and to the east to North-east Siberia, while southwards it ranges to the Pyrenees, North Italy, North Caucasus, the Tian Shan Mountains, and Peking. It is found as high as 69° N. lat. In some localities it is met with a little above the sea-level, while in Central Asia it ranges to 10,000 feet.

Changes of Plumage.—During the heavy autumn moult, which takes place in July and August, when the males are entirely devoid of tails and generally incapable of flying more than a few yards at most, a temporary protective plumage, like that of the female, clothes the head and neck, and the throat becomes more or less white. The object of this change is obvious, for the black head and neck of the male are conspicuous objects among the heather and rushes, but the rufous-buff feathers, with their black bars and marks, harmonise perfectly with these surroundings and enable the defenceless birds to escape the observation of their enemies. The barred feathers of the head and neck are not cast and replaced by black, till the rest of the plumage has been renewed, and the bird is once more able to fly.

The *young male*, unlike the Caucasian Black Grouse, attains the black adult plumage at the first autumn moult, and by November resembles the old male, but some of the finely mottled shoulder-feathers and inner flight-feathers of the first plumage are generally retained till the second season, and the outer tail-feathers are shorter and less beautifully curved.

Females that have become barren from age or accident commonly assume the male plumage to a greater or less extent, some examples having much black in the plumage and a very well-developed forked black tail, each feather being prettily edged with white. One peculiarity of these birds is the colour of the throat, which in the most fully plumaged examples is pure white.

The only time when the throat of the male is white, or partially so, is during the short period when the temporary hen-like plumage covers the head and neck. At that season the throat becomes white or thickly spotted with that colour. No doubt this is the source whence the pure white throat of the barren female is derived.

Habits.—Pine- and birch-forests are the true home of this bird, and though, when feeding, it may often be met with on the open moors or in the stubble-fields at a considerable distance from any covert, it is truly a denizen of the woods, and passes the greater part of its existence on the branches, where, unlike the Red Grouse, it is perfectly at home. Black Grouse, like other Game-Birds, are extremely partial to grain, and in some parts of Scotland, where they are still numerous, frequent the stubble-fields in enormous flocks, generally in the early morning and towards evening. They are polygamous— that is to say, one male pairs with many females, and generally towards the end of March or beginning of April the pairing-season commences, when the cocks are in the habit of repairing at dawn and sunset to some particular spot to display their charms to the females and give battle to their rivals.

The extraordinary pantomime gone through by each male as he struts round the arena, generally an open patch of ground worn nearly bare by constant traffic, is most entertaining to observe. With drooping wings, outspread tail, and many other curious antics, accompanied by an occasional spring into the air, he attempts to secure the goodwill of the ladies, and when two birds meet, a slight skirmish, in which a few feathers are lost, takes place. As a rule, no serious fights, such as one sees between Red Grouse, occur, merely a "round with the gloves," to entertain the ladies of the harem; but occasionally, when two old rivals chance to meet, a furious "set-to" may be witnessed, the fight lasting till one or both birds are thoroughly exhausted, bleeding and torn. These strange entertainments last till the females—or "Grey-hens" as they are called—have laid all their eggs and commenced to sit, when the males are seen no more, the hatching of the eggs and rearing of the young being exclusively the task of the females.

Hybrids between the Black-cock and female Capercailzie (so called *Tetrao medius*) are not uncommon, and it occasionally crosses with the Red Grouse, Willow Grouse, and more rarely with the Pheasant and Hazel-Hen.

Nest.—A slight hollow in the ground, scratched out and with little lining; usually well concealed.

Eggs.—Generally six to ten in number. Buff spotted with rich brown. Average measurements, 2 inches by 1·4.

II. THE CAUCASIAN BLACK GROUSE. LYRURUS MLOKOSIEWICZI.

Tetrao mlokosiewiczi, Tacz. P. Z. S. 1875, p. 266, woodcuts; Dresser, B. Europe, v. p. 219, pl. 488 (1876).
Tetrao acatoptricus, Radde, Orn. Caucas. p. 358 (1884); id. J. f. O. 1885, p. 79.
Lyrurus mlokosiewiczi, Ogilvie-Grant, Cat. B. Brit. Mus. xxii. p. 58 (1893).

Adult Male.—Plumage *entirely* black, including the under tail-coverts. Total length, 20 inches; wing, 8; tail, 8·2; tarsus, 2.

Adult Female.—Plumage mostly rufous and buff, barred with black, but the black bars and markings on the breast are much finer than in the female of *L. tetrix*, and the tail is longer. Total length, 16·6 inches; wing, 7·7; tail, 5·5; tarsus, 1·85.

Range.—This species is only found in the Caucasian Mountains.

Changes of Plumage.—The young males are peculiar in retaining a hen-like plumage throughout the first year, and probably till the second moult, thus differing entirely from the young males of *L. tetrix*, which attain their black plumage at the first autumn moult, and by December closely resemble their male parent.

THE CAPERCAILZIES. GENUS TETRAO.

Tetrao, Linn. S. N. i. p. 273 (1766).

Type, *T. urogallus*, Linn.

The members of this genus are all birds of large size, and, like the Black Grouse, have the tail composed of eighteen feathers, but are distinguished by having the middle pair of feathers much longer than the outer pair, which produces a rounded or wedge-shaped appearance when the tail is spread. There are no elongate tufts of feathers on each side of the neck, nor inflatable air-sacs in the male, and the outer flight-feathers are not attenuated or sickle-shaped.

I. THE CAPERCAILZIE. TETRAO UROGALLUS.

Tetrao urogallus, Linn. S. N. i. p. 273 (1766); Meyer, Unser Auer-, Rackel- und Birkwild, &c. pp. 1-15, pls. 1-3 (1887); Millais, Game-Birds, pp. 1-20, pls. and woodcuts (1892); Ogilvie-Grant, Cat. B. Brit. Mus. xxii. p. 60 (1893).

Adult Male.—Above dark grey, shading into reddish-brown on the wings and finely mottled with black; a metallic green band across the chest, and the throat glossed with the same colour. Middle of the back not barred with black; the shoulder-feathers not tipped with white; and the breast and belly black, a few feathers in the middle being tipped with white. Total length, 35 inches; wing, 14·6; tail, 12·3; tarsus, 2·8.

Adult Female.—Middle of the back rufous and buff, strongly barred with black; breast and belly buff or whitish-buff, barred with black; general colour of the plumage darker than in *T. uralensis*, the white tips to the scapulars being narrower. Total length, 25 inches; wing, 11·7; tail, 7.3; tarsus, 2·1.

Younger males resemble the adult, but are smaller, and the white band across the tail is wanting.

Nestling.—Very similar to that of *L. tetrix*.

Range.—The pine-forests of Europe and Northern and Central Asia, extending in the west to Scotland, in the east as far as Lake Baikal, and southwards to the Pyrenees, Alps, Carpathians, North-east Turkestan, and the Altai Mountains.

Habits.—The Capercailzie is an inhabitant of the pine-forests, and spends the greater part of its time among the branches, feeding on the tender shoots of spruce and larch; but it is also extremely fond of various ground-fruits, in search of which it may not unfrequently be found in comparatively open country at a considerable distance from the fir-woods. Like other Game-Birds, it also shows a great partiality for grain, visiting the stubble-fields in fine weather. On the ground, the movements of Capercailzie are slow and dignified, and when wounded, being incapable of running at any great pace, they seldom move far from where they fall, usually seeking concealment by hiding, at which, in spite of their size, they are great adepts. The weight of the old male averages from nine to twelve pounds, but, notwithstanding his bulk, the flight is easy and almost noiseless, though remarkably steady and rapid. It is astonishing how closely one of these great birds can glide past without its presence being detected, unless one happens to catch sight of it. The habits are somewhat similar to those of the Black Game, but the meeting-place of the males is generally some particular pine-tree known as the "laking-place." Here, in the month of April, the male may be seen at dawn and sunset, where, with outstretched neck, drooping wings, and tail erected and spread like a fan, he utters his "spel," or love-song. This consists of three notes, each being several times repeated, and towards the end of the song he works himself up into such a state of blind excitement that, careless of the surrounding objects, he heeds not the stealthy approach of the "sportsman," who takes advantage of these moments of ecstacy, and gradually gets within shooting distance. Thus on the Continent many a fine old cock is done to death, for only the older birds "spel," the younger and weaker cocks being driven from the field. Tremendous fights take place for the sovereignty of each harem, and both combatants may sometimes be captured, having fought till they are so completely exhausted that they are unable to escape.

The flesh of old birds has a strong flavour of turpentine, and, being extremely bitter, is unfit for the table, but young birds are often palatable enough, if properly cooked.

Hybrids.—The female Capercailzie, as already mentioned, frequently crosses with the Black-cock, and the *male hybrid* is a remarkably handsome bird, with a fine purplish gloss on the breast and a forked tail, but the latter is much less curved than that of the male parent. The *female hybrid* is much more difficult to distinguish, and may easily be mistaken for a large Grey-hen or small female Capercailzie; but there is an infallible means of distinguishing the three to be found in the comparative length of the middle tail-feathers and under tail-coverts. In the Grey-hen the tail is forked, the outer feathers being much the longest, and the under tail-coverts extend considerably beyond the middle pair. The female Capercailzie has the tail rounded, the middle pair of feathers being much longer than the outer, and the under tail-coverts do not extend nearly to the end of the middle pair, while in the female hybrid the tail is nearly square, the feathers being all of about the same length, and the under tail-coverts are much shorter than the middle pair. These hybrids have many names, such as *Tetrao hybridus*, &c. The best work on the subject is Dr. A. B. Meyer's volume quoted above.

The Capercailzie has also been known to cross with the Pheasant and Willow Grouse.

Nest and Eggs.—Very similar to those of the Black Grouse, but the eggs are somewhat larger than those of the latter bird. Average measurements, 2·2 by 1·6 inches.

SUB-SP. *a.* THE URAL CAPERCAILZIE. TETRAO URALENSIS.

Tetrao uralensis, Nazarov, Bull. Mosc. 1886, p. 365; Ogilvie Grant, Cat. B. Brit. Mus. xxii. p. 65 (1893).

(*Plate V.*)

Adult Male.—Similar to *T. urogallus*, but the mantle and back grey, finely mottled with black; wings and shoulder-feathers light reddish-brown, the latter not tipped with white; general colour of the upper-parts much paler than in *T. urogallus*; breast and belly mostly white.

Adult Female.—Mantle pale rufous and buff, strongly barred with black; the breast and belly buff or whitish-buff, barred with black; and the general colour of the plumage paler than in *T. urogallus*, the white tips to the shoulder-feathers being much wider.

Range.—The Ural Mountains.

Although at first sight this splendid Capercailzie, by far the handsomest of the genus, appears to be remarkably distinct from typical examples of *T. urogallus* from Norway and Sweden, I have examined numerous examples in intermediate stages of plumage between the dark Scandinavian bird and the light-coloured Ural form. These intermediate birds come into the London market in considerable numbers, and are believed to be imported from some of the more southern

provinces of Russia, but, so far, I have been unable to ascertain the exact locality whence they are obtained. It must be added that, though some of these intermediate birds have much white on the breast and belly, and are altogether lighter than Western European examples, the Ural birds are so very much paler, and show no trace of variation among themselves, that they may be fairly considered at present as representing a well-marked geographical sub-species, though most probably future investigations will show that they completely intergrade with typical western and eastern forms.

II. THE SLENDER-BILLED CAPERCAILZIE. TETRAO PARVIROSTRIS.

Tetrao urogalloides, Middend. (*nec* Nilss.*), Sibir. Reise, ii. pt. ii. p. 195, pl. xviii. (1851); Elliot, Mon. Tetraon. pl. vi. (1865).

Tetrao parvirostris, Bonap. C. R. xlii. p. 880 (1856); Ogilvie-Grant, Cat. B. Brit. Mus. xxii. p. 66 (1893).

Adult Male.—Mantle brownish-black, not barred, and the shoulder-feathers less widely tipped with white, the white tips forming *an interrupted line of white spots*. Total length, 35 inches; wing, 15; tail, 14·8; tarsus, 2·7.

Adult Female.—Mantle strongly barred with black; the breast and belly black, barred with buff and tipped with white; and the white spots on the shoulder-feathers forming *an interrupted line of white spots*. Total length, 25 inches; wing, 11·6; tail, 7·7; tarsus, 1·7.

Range.—This species takes the place of *T. urogallus* in the pine-forests of the north east of Siberia to the east of Lake Baikal, and is also found in the island of Saghalien, but not in Kamtschatka.

---

* This name was previously used by Nilsson for the hybrid between the Black Grouse and Capercailzie.

III. THE KAMTSCHATKAN CAPERCAILZIE. TETRAO KAMTSCHATICUS.

Tetrao kamtschaticus, Kittl. Reise Kamtschatka, ii. p. 353, woodcut (1858); Ogilvie-Grant, Cat. B. Brit. Mus. xxii. p. 67 (1893).

Adult Male.—Like *T. parvirostris*, but smaller; mantle not barred with black; shoulder-feathers broadly tipped with white, forming a *continuous white band* down each side of the back. Total length, 30 inches; wing, 14·1; tail, 11·0; tarsus, 2·7.

Adult Female.—Mantle strongly barred with black; breast and belly black, barred with buff and tipped with white; the white tips of the shoulder-feathers form a *continuous white band*, as in the male. Total length, 22 inches; wing, 11·1; tail, 6·4; tarsus, 2.

Range.—This species is only known to occur in Kamtschatka, where it replaces *T. parvirostris*.

THE CANADIAN GROUSE. GENUS CANACHITES.

Canachites, Stejn. P. U. S. Nat. Mus. viii. p. 409 (1885).

Type, *C. canadensis* (Linn.).

Toes naked and pectinate on the sides; tail fairly long and rounded, composed of sixteen feathers, the outer pair being not much shorter than the middle pair; no elongate tufts of feathers on each side of the neck, and the outer flight-feathers not attenuated or sickle-shaped.

This genus includes only two small North American species of about the size of the Common Partridge of Europe.

I. THE CANADA GROUSE. CANACHITES CANADENSIS.

Tetrao canadensis, Linn. S. N. i. p. 274 (1766); Audub. Orn.
  Biogr. ii. p. 437, pl. clxxvi. (1834); v. p. 563 (1839).
Canace canadensis, Reichenb. Av. Syst. Nat. p. xxix. (1851);
  Elliot, Monogr. Tetraon. pl. ix. (1865).

Canachites canadensis, Stejn. P. U. S. Nat. Mus. viii. p. 409 (1885); Ogilvie-Grant, Cat. B. Brit. Mus. xxii. p. 69 (1893).
Dendragapus canadensis, Bendire, Life Hist. N. Am. B. p. 51, pl. 1, figs. 20-23 (1892).

Adult Male.—General colour above black, barred with brownish-grey; upper tail-coverts edged or tipped with *grey*; the tail with a well-marked terminal rufous band; chin, throat, and most of under-parts black. Total length, 15·5 inches; wing, 6·8; tail, 4·8; tarsus, 1·4.

Adult Female.—May be distinguished from the male in having the chin and throat rufous, spotted with black; the neck and chest black, barred with rust-colour, and the rest of the under-parts the same, but tipped with white. Total length, 14·2 inches; wing, 6·8; tail, 4; tarsus, 1·4.

Range.—This Grouse inhabits the northern parts of North America, ranging westwards to the east side of the Rocky Mountains, eastwards to New England and New York, northwards to Alaska, and south to Minnesota.

Habits.—The favourite haunts of this handsome little species are dense thickets and evergreen woods. Its food consists largely of the tender spruce buds and needles, varied in summer with berries of various kinds. The pairing-season commences in the end of April or early in May, the eggs being laid in the latter part of May or beginning of June. Unlike the Capercailzies and Black Grouse, these birds are monogamous, and there is good reason to believe that some retain their mates for more than one season, isolated pairs being often found together in the middle of winter. During the breeding-season the male has a peculiar habit of drumming, which has been described as follows: "After strutting back and forth for a few minutes, the male flew straight up as high as the surrounding trees, about fourteen feet; here he remained stationary an instant, and while on suspended wing did the drumming with the wings, resembling distant thunder, meanwhile dropping down slowly to the spot from whence he started, to repeat the same thing over and over again." Capt. Bendire gives another description of the drumming: "The Canada Grouse performs its 'drumming' upon the trunk of a standing tree of rather small size, preferably one that is inclined from the perpendicular, and in the following manner. Commencing near the base of the tree selected, the bird flutters upward with somewhat slow progress, but rapidly beating wings, which produce the drumming sound. Having thus ascended fifteen or twenty feet, it glides quietly on wing to the ground and repeats the manœuvre. Favourite places are resorted to habitually, and these 'drumming trees' are well-known to observant woodsmen. I have seen one so well worn upon the bark as to lead to the belief that it had been used for this purpose for many years."

Eggs.—Seven to thirteen, sometimes more. Similar to those of *L. tetrix*, but smaller, and the ground-colour sometimes reddish-buff; the markings, also, are generally heavier, some of the spots being confluent and forming blotches. Average measurements, 1·75 inch by 1·25.

II. FRANKLIN'S GROUSE. CANACHITES FRANKLINI.

Tetrao franklini, Dougl. Trans. Linn. Soc. xvi. p. 139 (1829);
   Swains. and Richards. Faun. Bor.-Amer. ii. p. 348, pl. lxi. (1831).
Canace franklini, Elliot, Proc. Acad. Philad. 1864, p. 23; id. Monogr. Tetraon. pl. x. (1865).
Dendragapus franklinii, Bendire, Life Hist. N. Am. B. p. 56 (1892).
Canachites franklini, Ogilvie-Grant, Cat. B. Brit. Mus. xxii. p. 71 (1893).

Adult Male.—Like the male of *C. canadensis*, but the upper tail-coverts tipped with white; no terminal rufous band to the tail; chin and throat black.

Adult Female.—Distinguished by having the chin and throat rufous, spotted with black.

Range.—The west side of the northern Rocky Mountains, extending westwards to the coast ranges.

Habits.—The habits of this species are apparently very similar to those of the Canada Grouse, which it replaces to the west of the Rocky Mountains. It is found in the almost inpenetrable and densely-timbered mountain ranges, generally, at an altitude of from 5,000 to 9,000 feet, in the neighbourhood of running water or swampy valleys. It is said to be a remarkably fearless and stupid bird, frequently allowing itself to be knocked off the trees with sticks or stones, and it is often caught by hand.

THE SHARP-WINGED GROUSE. GENUS FALCIPENNIS.

Falcipennis, Elliot, P. Ac. Philad. 1864, p. 23.

Type, *F. falcipennis* (Hartl.).

Toes naked and pectinate along the sides. Tail moderately long and rounded, composed of sixteen feathers. *The outer flight-feathers attenuated and sickle-shaped.*

1. THE SHARP-WINGED GROUSE. FALCIPENNIS FALCIPENNIS.

Tetrao falcipennis, Hartl. J. f. O. 1855, p. 39.
Falcipennis hartlaubi, Elliot, Proc. Acad. Philad. 1864, p. 23; id. Monogr. Tetraon. pl. xi. (1865).
Falcipennis falcipennis, Ogilvie-Grant, Cat. B. Brit. Mus. xxii. p. 72 (1893).

Adult Male.—Chest uniform smoky-black. Total length, 16·3 inches; wing, 7·2; tail, 4·7; tarsus, 1·4.

Adult Female.—Chest black, barred with buff. Total length, 14·7 inches; wing, 7·2; tail, 4·3; tarsus, 1·4.

Range.—North-eastern Siberia, Kamtschatka, and Saghalien Island.

In general appearance and size this species resembles the

Canada Grouse (*C. canadensis*), and may be regarded as the representative form of that species in the Old World.

THE AMERICAN CAPERCAILZIES. GENUS DENDRAGAPUS.

Dendragapus, Elliot, Proc. Acad. Philad. 1864, p. 23.

Type, *D. obscurus* (Say).

Toes naked and pectinate along the sides. Tail long, composed of twenty feathers sub-equal in length. The male is provided with *an inflatable air-sac on each side of the neck*, but there are no elongate tufts of feathers, nor are the outer flight-feathers attenuated or sickle-shaped.

This genus includes three rather large forms, about the size of a Black Grouse, but, unlike these birds and the True Capercailzies, the American Capercailzie seems to pair with one female only.

1. DUSKY CAPERCAILZIE. DENDRAGAPUS OBSCURUS.

Tetrao obscurus, Say, in Long's Exped. Rocky Mts. ii. p. 14 (1823); Bonap. Amer. Orn. iii. p. 27, pl. xviii. (1828).

Dendragapus obscurus, Elliot, Proc. Acad. Philad. 1864, p. 23, and Monogr. Tetraon. pl. vii. (1865); Bendire, Life Hist. N. Am. B. p. 41 (1892); Ogilvie-Grant, Cat. B. Brit. Mus. xxii. p. 74 (1893).

Adult Male.—General colour above smoky-black, mixed with brownish-buff, below grey; chest and breast *not* barred and marked with buff; tail somewhat rounded, with a wide terminal grey band varying in width on the *middle* feathers from 1 to 1·5 inch. Total length, 19·5 inches; wing, 10; tail, 6·7; tarsus, 1·7.

Adult Female.—Chest and breast barred and marked with buff; tail with a wide grey terminal band, about 0·8 inch in width, on the *outermost* feathers. Total length, 17 inches; wing, 8·6; tail, 5·9; tarsus, 1·6.

Range.—Southern Rocky Mountains, extending in the west to Wahsatch, in the south to New Mexico and Arizona, and north to the South Pass.

In South-eastern Idaho the Dusky Capercailzie is said to intergrade with the darker and more northern form, *D. fuliginosus*.

Habits.—This and the two allied forms are perhaps the finest, and, with the exception of the Sage Cock, the largest of the American Grouse. In the males of all three the general colour of the plumage is smoky-black, and hence the present species is often known as "Blue Grouse" as well as "Pine Grouse" and "Pine Hen." From Mr. Gale's interesting notes published in Captain Bendire's excellent work, "Life Histories of North American Birds," so often alluded to in these pages, the following account is taken: "Here in Colorado the Dusky Grouse ranges from an altitude of about 7,000 feet to the timber-line. Having once selected a place to raise a brood they do not stray far from the neighbourhood. Water at no great distance is always kept in view. The lower gulches and side hills are mostly chosen for their summer homes. During the mating-season, if you are anywhere near the haunts of a pair, you will surely hear the male and most likely see him. He may interview you on foot, strutting along before you, in short hurried tacks alternating from right to left, with widespread tail tipped forward, head drawn in and back, and wings dragging along the ground, much in the style of a Turkey-gobbler. At other times you may hear his mimic thunder overhead again and again, in his flight from tree to tree. As you walk along, he leads, and this reconnoitring on his part, if you are not familiar with it, may cause you to suppose that the trees are alive with these Grouse. He then takes his stand upon a rock, stump, or log, and, in the manner already described, distends the lower part of his neck, opens his frill of white, edged with the darker feather tips, showing in the centre a pink narrow line describing somewhat the segment of a circle, then with very little apparent motion he performs his growling or groaning, I don't know which to call it, having the strange peculiarity of seeming quite distant when quite near, and near when distant, in fact appearing to come from every direction but the true one. . . . As near as I can judge by meeting with the young broods, these birds nest at the lowest points about May 15, at the highest about the beginning of June. The number of chicks seen by me in a brood ranged from three to eight. . . . In a single instance only, with a brood about ten days old, have I noticed the presence of both parents. Perched upon a fallen tree, the male seemed to be on the look-out, while the female and young were feeding close by. This seeming indifference of the male while the brood is very young, allowing his mate to protect them, if he really is always near at hand, looks very strange, and yet it may be the case, since he is generally with the covey when the young are well-grown."

Eggs.—Pale cream-colour to creamy-buff, equally marked all over with rather small rounded spots and dots of chestnut brown. Average measurements, 1·9 by 1·4 inch.

SUB-SP. *a*. THE SOOTY CAPERCAILZIE. DENDRAGAPUS FULIGINOSUS.

Canace obscurus, var. *fuliginosus*, Baird, Brewer and Ridgw. N. Amer. B. iii. p. 425 (1874).

Dendragapus obscurus fuliginosus, Bendire, Life Hist. N. Am. B. p. 43, pl. i. figs. 16-19 (1892).

Dendragapus fuliginosus, Ogilvie-Grant, Cat. B. Brit. Mus. xxii. p. 75 (1893).

Adult Male.—Distinguished from *D. obscurus* by having the grey band across the tip of the tail narrower, *less than an inch wide on the middle feathers*, and by the somewhat darker and more uniform plumage, with much fewer buff markings on the upper-parts, especially on the wing-coverts.

Adult Female.—Can only be recognised from the female of

D. obscurus by the narrower grey band across the end of the tail.

This is merely a sub-species or race of the Dusky Capercailzie.

Range.—North-western Rocky Mountains near the Pacific Coast, from California to Sitka and Alaska.

II. RICHARDSON'S CAPERCAILZIE.   DENDRAGAPUS RICHARDSONI.

Tetrao richardsonii, Dougl. Trans. Linn. Soc. xvi. p. 140 (1829); Wilson, Illustr. Zool. pls. xxx. xxxi. (1831).
Dendragapus richardsoni, Elliot, Monogr. Tetraon. pl. viii. (1865); Ogilvie-Grant, Cat. B. Brit. Mus. xxii. p. 76 (1893).
Dendragapus obscurus richardsonii, Bendire, Life Hist. N. Am. B. p. 50 (1892).

Adult Male.—Easily recognised from the two forms previously mentioned by having the tail *uniform black*, without a grey band across the extremity. The tail is also squarer in shape, the outer feathers being slightly longer than the middle pair.

Adult Female.—Resembles the females of the Dusky and Sooty Capercailzie, but has no grey band across the tail, though the feathers are usually *margined with grey at the tip*.

Range.—Eastern slopes of the northern Rocky Mountains, from Montana northwards into British America.

In Northern Wyoming and the eastern parts of Central Idaho this species is said to intergrade with *D. obscurus*, and in North-eastern Idaho and Western Montana with *D. fuliginosus*.

THE PINNATED GROUSE.   GENUS TYMPANUCHUS.

Tympanuchus, Glog. Hand. u. Hilfsb. p. 396 (1842).

Type, *T. cupido* (Linn.).

Toes naked and pectinate along the sides. Tail rather short and rounded and composed of eighteen feathers, the outer pair about two-thirds the length of the middle pair.

The males have *an elongated tuft of feathers* and an inflatable air-sac on each side of the neck.

I. THE PRAIRIE HEN. TYMPANUCHUS AMERICANUS.

Cupidonia americana, Reichenb. Av. Syst. Nat. p. xxix. (1852).
Tetrao cupido, Wils. (*nec* Linn.), Am. Orn. iii. p. 104, pl. 27, fig. 1 (1811).
Cupidonia cupido, Baird (*nec* Linn.), B. N. Amer. p. 628 (1860) [part]; Elliot, Monogr. Tetraon. pl. xvi. (1865).
Tympanuchus americanus, Ridgw. Auk. iii. p. 132 (1886); Bendire, Life Hist. N. Am. B. p. 88 (1892); Ogilvie-Grant, Cat. B. Brit. Mus. xxii. p. 78 (1893).

Adult Male.—Above barred with rufous- or brownish-buff and black; below barred with brownish-black and white. Feathers of the neck-tufts much produced, about 3 inches in length, the longer ones being *parallel-edged*, with *rounded* or *truncate* extremities; chest-feathers white, with *two brown bars as wide as the white interspaces;* outer tail-feathers black, narrowly tipped with white. Total length, 16·5 inches; wing, 9; tail, 3·9; tarsus, 1·9.

Adult Female.—Differs in having the neck-tufts short; the outer tail-feathers barred with rufous-buff. Measurements usually a trifle less than those of the male.

Range.—This species inhabits the prairies of the Mississippi Valley, extending northwards to southern Manitoba and Wisconsin, south to Louisiania and Texas, east to Indiana, Kentucky, and North-western Ohio, and west to Indian Territory, Kansas, Nebraska, and Dakota.

Habits.—This species is a resident throughout the greater part of its range, but it seems that in Iowa a regular though local migration takes place.

As soon as severe weather sets in, large flocks of these birds leave the northern prairies and go south to winter in Northern Missouri and Southern Iowa, the migration varying in bulk with the severity of the winter. The curious feature of this migration is, that only the *females* are believed to change their quarters. Writing from Minnesota, Mr. Miller says: "The females in this latitude migrate south in the fall and come back in spring, about one or two days after the first Ducks, and they keep coming in flocks of from ten to thirty for about three days, all flying north. The Grouse that stay all the winter are males."

Captain Bendire publishes the following amusing and interesting account of the love-making of this species: "Early in the morning you may see them assemble in parties, from a dozen to fifty together, on some high dry knolls, where the grass is short, and their goings on would make you laugh. The cock birds have a loose patch of naked yellow skin on each side of the neck just below the head, and above these on either side, just where the head joins the neck, are a few long black feathers, which ordinarily lay backward on the neck, but which, when excited, they can pitch straight forward. Those yellow naked patches on either side of the neck cover sacs which they can blow up like a bladder whenever they choose. These are their ornaments, which they display to the best advantage before the gentler sex at these love-feasts. This they do by blowing up these air-sacs till they look like two ripe oranges, on each side of the neck, projecting their long black ears right forward, ruffling up all the feathers of the body till they stand out straight, and dropping their wings on the ground like a Turkey cock. . . .

"'Then it is that the proud cock, in order to complete his triumph, will rush forward at its best speed for two or three rods through the midst of the love-sick damsels, pouring out as he goes a booming noise, almost a hoarse roar, only more subdued, which may be heard for at least two miles in the still morning air. This heavy booming sound is by no means harsh or unpleasant, on the contrary it is soft and even harmonious. When standing in the open prairie at early dawn listening to hundreds of different voices, pitched on different keys, coming from every direction and from various distances, the listener is rather soothed than excited. If this sound is heavier than the deep key-notes of a large organ, it is much softer, though vastly more powerful, and may be heard at a much greater distance. One who has heard such a concert can never after mistake or forget it.

"Every few minutes this display is repeated. I have seen not only one, but more than twenty cocks going through this funny operation at once, but then they seem careful not to run against each other, for they have not yet got to the fighting point. After a little while the lady birds begin to show an interest in the proceedings, by moving about quickly a few yards at a time, and then standing still a short time. When these actions are continued by a large number of birds at a time, it presents a funny sight, and you can easily think they are moving to the measure of music.

"The party breaks up when the sun is half an hour high, to be repeated the next morning and every morning for a week or two before all make satisfactory matches. It is toward the latter part of the love-season that the fighting takes place among the cocks, probably by two who have fallen in love with the same sweetheart, whose modesty prevents her from selecting between them." According to Bendire, immense numbers of nests of this species "are annually destroyed, either by fire in dry seasons, or water during wet ones. . . . On the prairies they generally select unburnt places to nest in, where the old grass is thick; others prefer the borders of large marshes, where, during a wet season, they are almost certain to be destroyed by water." Many nests and eggs are also yearly ploughed up, as cultivated fields and meadows are often selected.

Nest.—A slight excavation in the ground, generally without any lining, but sometimes lined when materials are available.

Eggs.—Eleven to fourteen in number, or even more. Ground-colour pale buff, olive-buff, or vinaceous, with very small, sometimes obsolete, dots of chestnut-brown.

II. THE HEATH HEN. TYMPANUCHUS CUPIDO.

Tetrao cupido, Linn. S. N. i. p. 274 (1766).
Cupidonia cupido, Brewst. Auk. ii. p. 82 (1885).
Tympanuchus cupido, Ridgw. P. U. S. Nat. Mus. viii. p. 355 (1885); Bendire, Life Hist. N. Am. B. p. 93 (1892); Ogilvie-Grant, Cat. B. Brit. Mus. xxii. p. 77 (1893).

Adult Male.— Similar to the foregoing species, but with fewer feathers in the neck-tufts; *the longer ones lanceolate and pointed.*

Adult Female.—Resembles the female of *T. americanus.*

This species is a smaller form, very closely allied to the Prairie Hen, but the male may apparently be distinguished by the above-mentioned characters.

Range.—Island of Martha's Vineyard, Massachusetts. It was formerly also found in Eastern Massachusetts, Connecticut, Long Island, New Jersey, and Pennsylvania, according to American records, but is now extinct in these localities.

Habits.—The habits of this bird are somewhat different from those of its western ally, for it is a woodland species, only met with in the scrubby tracts of oak, and feeding largely on acorns, though it may occasionally be seen in the open picking up grain and clover-leaves. The area inhabited by the remaining colony of these birds covers about forty square miles, and over this extremely limited range they are comparatively numerous, being now strictly protected by law.

III. THE LESSER PRAIRIE HEN. TYMPANUCHUS PALLIDICINCTUS.

Cupidonia cupido, var. *pallidicinctus*, Ridgw. in Baird, Brewer. & Ridgw. N. Amer. B. iii. p. 446 (1874).

Tympanuchus pallidicinctus, Ridgw. P. U. S. Nat. Mus. viii.
p. 355 (1885); Bendire, Life Hist. N. Am. B. p. 96
(1892); Ogilvie-Grant, Cat. B. Brit. Mus. xxii. p. 80
(1893).

Adult Male.—Distinguished by the longer feathers of the neck-tufts, these being *parallel-edged* and *square-tipped*; chest-feathers white, with *three brown bars, narrower than the white interspaces*. Total length, 15 inches; wing, 8·3; tail, 3·5; tarsus, 1·6.

Adult Female.—Differs from the *male* in having the neck-tufts much shorter, and, as in the female of *T. americanus*, the outer tail-feathers are barred with buff. Measurements a trifle less than those of the male.

Range.—South-western Kansas, Western Indian Territory, Western, and probably Southern, Texas.

The range of this smaller and paler-coloured species is still imperfectly known, but its nesting habits appear to be very similar to those of *T. americanus*.

THE SAGE GROUSE. GENUS CENTROCERCUS.

Centrocercus, Swains. Faun. Bor.-Amer. ii. pp. 342, 496 (1831).

Type, *C. urophasianus* (Bonap.).

Toes naked and pectinate along the sides; tail long and Pheasant-like, composed of twenty wedge-shaped pointed feathers, the outer pair being less than two-thirds the length of the middle pair. The males have an inflatable air-sac on each side of the neck. Only one species is known.

1. THE SAGE GROUSE. CENTROCERCUS UROPHASIANUS.

Tetrao urophasianus, Bonap. Zool. Journ. iii. p. 213 (1828);
id. Amer. Orn. iii. p. 55, pl. xxi. fig. 1 (1828).
Tetrao (Centrocercus) urophasianus, Swains. & Richards. Faun.
Bor.-Amer. ii. p. 358, pl. 58 (1831).

PLATE VI

SAGE GROUSE

Centrocercus urophasianus, Jard. Nat. Libr. Orn. iv. p. 140, pl. xvii. (1834); Elliot, Monogr. Tetraon. pl. xiii. (1865). Bendire, Life Hist. N. Am. B. p. 106, pl. iii. figs. 11-12 (1892); Ogilvie-Grant, Cat. B. Brit. Mus. xxii. p. 81 (1893).

(*Plate VI.*)

Adult Male.—General colour above blackish, marked and mottled with buff; breast and belly mostly black; the chin and throat white, spotted with black; otherwise very similar to the female, though much larger, attaining a weight of eight pounds. Total length, 28 inches; wing, 12·5; tail, 12; tarsus, 2·2.

Adult Female.—Has the chin and throat white, and is much smaller, rarely weighing more than five pounds. Total length, 22 inches; wing, 10·8; tail, 6·5; tarsus, 1·9.

Range.—The sage-brush plains of the Rocky Mountain plateau, extending northwards to British America and south to New Mexico, South California, Utah, and Nevada.

The Sage Grouse is the largest species of its kind found in the New World, and is generally resident in those States where it occurs, but, like the Prairie Hen, it is also partially migratory in some parts of its range. As its name implies, this bird is seldom found far from the tracts of sage-brush (*Artemisia*), the leaves of which form its principal food, at least during the winter. As Captain Bendire explains, though the Grouse breed abundantly on the higher altitudes of about 6,000 feet, the bushes at that elevation become covered with snow in winter, and the birds are then driven down to the valleys in search of food, and thus a partial migration takes place in the beginning of winter and spring. In summer the food is varied with wild peas, seeds, grain, and insects, and the flesh is then excellent, provided that the birds are drawn as soon as they are shot, though in winter, when sage-leaves form the principal or only diet, they are unfit for the table. In the beginning of March the males pay their court to the females, and with distended air-sacs almost hiding the head, outspread tail, and trailing wings, they strut slowly about before the females, uttering meanwhile low gutteral sounds. The males do not take any part in the incubation, and remain apart till the young are grown. In autumn the birds pack, and may then be seen in large numbers.

The stomach of the Sage Grouse differs from that of all other Game Birds in being soft, very different from the muscular gizzard found in all the allied forms.

Nest.—A slight hole scratched in the ground, with little or no lining; generally placed under the shelter of a small sage-bush.

Eggs.—Seven to nine in number, sometimes more, as many as seventeen having been found in a nest. Colour varying from olive-buff to greenish-brown, rather heavily dotted all over with well-defined chocolate-brown spots. Average measurements, 2·2 inches by 1·5.

THE SHARP-TAILED GROUSE. GENUS PEDIŒCETES.

Pediocætes, Baird, Rep. Expl. & Surv. ix. pt. 2, Zool. p. 625 (1858).

Type, *P. phasianellus* (Linn.).

Toes naked and pectinate along the sides. Tail rather long and wedge-shaped, composed of eighteen feathers, the middle pair being more than twice as long as the outer pair.

The males are provided with an inflatable air-sac on each side of the neck, but have no elongate neck-tufts.

1. THE NORTHERN SHARP-TAILED GROUSE. PEDIŒCETES PHASIANELLUS.

Tetrao phasianellus, Linn. S. N. i. p. 273 (1766).
Pediocætes kennicotti, Suckl. P. Ac. Philad. 1861, pp. 334, 361.
Pediocætes phasianellus, Elliot, P. Ac. Philad. 1862, p. 403;
  Bendire, Life Hist. N. Am. B. p. 97, pl. iii. figs. 3-5 (1892);
  Ogilvie-Grant, Cat. B. Brit. Mus. xxii. p. 82 (1893).

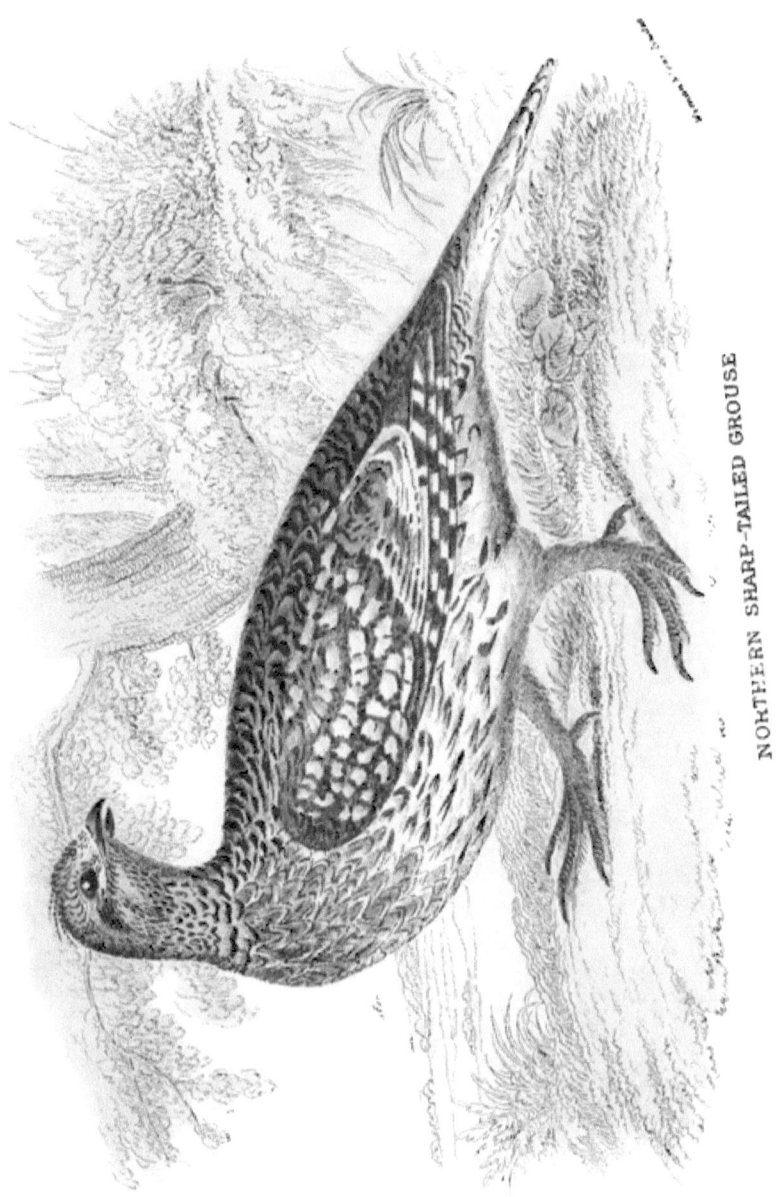

NORTHERN SHARP-TAILED GROUSE

Pediœcetes phasianellus, Blakiston, Ibis, 1862, p. 8.
Pediœcætes phasianellus, Elliot, Monogr. Tetraon. pl. xv. (1865).
(*Plate VII.*)

Adult Male and Female.—Distinguished from the smaller and more southern form, *P. columbianus*, by having the general colour above dark, the black on the upper-parts predominating over the rufous, buff, and white markings; the feathers of the chest black, with a white heart-shaped patch in the middle and a white fringe round the margin. Male measures: Total length, 16·8 inches; wing, 8·4; tail, 4·6; tarsus, 1·5. Female somewhat smaller.

Range.—Interior of British America, extending north to Fort Simpson, south to Lake Winnipeg and the north shore of Lake Superior, east to Hudson Bay, and west to the Rocky Mountains.

Habits.—The habits of this form appear to be very similar to those of its more southern representative. It inhabits the wooded districts and borders of the tundras near the lakes.

Eggs.—Seven to fourteen in number. Fawn brown, chocolate, or tawny, covered with small well-marked reddish-brown spots and dots. Average measurements, 1·8 by 1·3 inch.

II. THE COLUMBIAN SHARP-TAILED GROUSE. PEDIŒCETES COLUMBIANUS.

Phasianus columbianus, Ord, Guthrie's Geogr. (2nd Amer. ed.), ii. p. 317 (1815).
Pediœcetes columbianus, Elliot, P. Ac. Philad. 1862, p. 403; Bendire, Life Hist. N. Am. B. p. 98, pl. iii. figs. 6-8 (1892); Ogilvie-Grant, Cat. B. Brit. Mus. xxii. p. 83 (1893).
Pediœcætes columbianus, Elliot, Monogr. Tetraon. pl. xiv. (1865).
Pediœcetes columbianus, Cooper, Orn. Calif. i. p. 532 (1870).
Pediœcetes phasianellus campestris, Ridgw. P. Biol. Soc. Wash. ii. p. 93 (1884); Bendire, Life Hist. N. Am. B. p. 98, pl. iii. figs. 6-8 (1892).

Adult Male and Female.—Distinguished from the northern form, *P. phasianellus*, by their smaller size, and by having the general colour above lighter, the rufous-buff and white markings predominating over the black; the feathers on the breast white, each with a concentric sub-marginal black band. Male measures: Total length, 15 inches; wing, 8·1; tail, 5; tarsus, 1·7. Female somewhat smaller.

Range.—Plains of the United States; extending north to Manitoba, east to Wisconsin and Northern Illinois, south to New Mexico, and west from Northern California west of the Rocky Mountains to Fort Yukon, Alaska.

Capt. Bendire publishes some interesting notes by Mr. Thompson on the habits of the Prairie Chicken, as it is commonly called, from which the following extract is taken: "After the disappearance of the snow, and the coming of warmer weather, the Chickens meet every morning at grey dawn in companies of from six to twenty, on some selected hillock or knoll, and indulge in what is called 'the dance.' This performance I have often watched, and it presents the most amusing spectacle I have yet witnessed in bird-life. At first the birds may be seen standing about in ordinary attitudes, when suddenly one of them lowers its head, spreads out its wings nearly horizontally and its tail perpendicularly, distends its air-sacs and erects its feathers, then rushes across the 'floor,' taking the shortest of steps, but stamping its feet so hard and rapidly, that the sound is like that of a kettledrum; at the same time it utters a sort of bubbling crow, which seems to come from the air-sacs, beats the air with its wings and ·tes its tail, so that it produces a loud, rustling noise, and thus contrives at once to make as extraordinary a spectacle of itself as possible. As soon as one commences, all join in, rattling, stamping, drumming, crowing, and dancing together furiously; louder and louder the noise, faster and faster the dance becomes, until at last, as they madly whirl about, the birds leap over each other in their excitement. After a brief spell the energy of the dancers begins to abate,

PLATE VII.

and shortly afterward they cease, and stand or move about very quietly, until they are again started by one of their number leading off."

Nest.—A hollow in the ground, generally lined with grass and well-concealed.

Eggs.—Less richly coloured than those of the northern form.

THE RUFFED GROUSE. GENUS BONASA.

Bonasa, Steph. in Shaw's Gen. Zool. xi. p. 298 (1819).

Type, *B. umbellus* (Linn.).

Tarsi only partially feathered, the lower part being entirely naked. Toes naked and pectinate along the sides. Tail composed of eighteen feathers, rather long and bluntly wedge-shaped, the outermost pair being nearly as long as the middle pair. Sexes similar in plumage. A frilled ruffle of fan-shaped feathers on each side of the neck. Only one North American species is known.

1. THE RUFFED GROUSE. BONASA UMBELLUS.

Tetrao umbellus, Linn. S. N. i. p. 275 (1766).
Tetrao togatus, Linn. S. N. i. p. 275 (1766).
Bonasa umbellus, Steph. in Shaw's Gen. Zool. xi. p. 300 (1819);
    Bendire, Life Hist. N. Am. B. p. 59, pl. ii. fig. 1 (1892);
    Ogilvie-Grant, Cat. B. Brit. Mus. xxii. pp. 85, 558 (1893).
Tetrao umbelloides, Dougl. Tr. Linn. Soc. xvi. p. 148 (1829).
Tetrao sabinii, Dougl. Tr. Linn. Soc. xvi. p. 137 (1829).
Bonasa umbellus togata, Bendire, Life Hist. N. Am. B. p. 64, pl. ii. fig. 2 (1892).
Bonasa umbellus umbelloides, Bendire, Life Hist. N. Am. B. p. 67, pl. ii. fig. 3 (1892).
Bonasa umbellus sabini, Bendire, Life Hist. N. Am. B. p. 68, pl. ii. fig. 4 (1892).

(*Plate VIII.*)

Characters and Range.—The plumage of *male* and *female* alike is subject to great climatic variation, some individuals having the general colour of the upper-parts rufous, and others mostly grey, while in a large percentage of examples every intermediate shade can be found. The dark barring on the feathers of the under-parts also varies greatly in intensity, being sometimes extremely faint, at others strongly marked and edged with lines of a deeper colour.

The various varieties have been classed by American ornithologists under four different names, and are regarded as distinct sub-species; but as the differences in colour are by no means entirely dependent on locality, and grade imperceptibly into one another, it seems useless to employ names which only apply to the more extreme forms of each type, while a large majority of specimens, representing every intermediate phase of colour and markings, may have either two or more names applied to them with equal correctness. According to the American ornithologists, the darker rufous variety (*Bonasa sabinii*) is mostly met with in the wooded countries between the western slopes of the Coast Range and the Pacific Ocean, where the rainfall is very heavy; and it occurs as far north as Sitka. In Alaska, the Central Rocky Mountains eastwards through British North America, and southwards to Utah and Colorado, a lighter-coloured grey form (*B. umbelloides*) is found on the high ground. A somewhat darker form, with the dusky breast-bars more defined, inhabits British Columbia, Washington, and Oregon, and extends eastwards through Canada to the mountains of New England. Lastly, in the Eastern United States, as far south as the mountains of North Alabama, the lighter rufous form (*B. umbellus*) is met with. In a series of skins from the State of New York alone, however, all these varieties are more or less perfectly represented, though perhaps not in their most typical forms, and I therefore consider it needless to employ more than one name for all the various phases of this polymorphic species. *Males* measure: Total length, 17 inches; wing, 7·3; tail, 6·5; tarsus, 1·6. The *female* is rather smaller.

Habits.—Capt. Bendire writes: "The Ruffed Grouse is partial to an undulating and hilly country, one well-wooded and covered with considerable undergrowth, interspersed here and there with cultivated fields and meadow lands. In the southern portions of its range, this bird is confined to the more mountainous and Alpine regions, being seldom found far away from such places, excepting in the late fall.

"As winter approaches, the coveys leave their feeding-grounds in the mountains, and repair to more congenial haunts along the edges of the neighbouring valleys."

Mr. Ernest E. Thompson, writing from Canada, says: "Every fieldman must be acquainted with the simulation of lameness, by which many birds decoy, or try to decoy, intruders from their nests. This is an invariable device of the Ruffed Grouse, and I have no doubt that it is quite successful with the natural foes of the bird; indeed, it is often so with man. A dog, as I have often seen, is certain to be misled and duped, and there is little doubt that a Mink, Skunk, Raccoon, Fox, Coyote, or Wolf, would fare no better. Imagine the effect of the bird's tactics on a prowling Fox. He has scented her as she sits, he is almost upon her, but she has been watching him, and suddenly with a loud 'whir' she springs up and tumbles a few yards before him. The suddenness and noise with which the bird appears, causes the Fox to be totally carried away; he forgets all his former experience, he never thinks of the eggs, his mind is filled with the thought of the wounded bird almost within his reach; a few more bounds and his meal will be secured. So he springs and springs, and very nearly catches her, and in his excitement he is led on and away, till finally the bird flies off, leaving him a quarter of a mile or more from the nest.

"If, instead of eggs, the Partridge has chicks, she does not await the coming of the enemy, but runs to meet and mislead him ere yet he is in the neighbourhood of the brood; she then leads him far away, and, returning by a circuitous route, gathers her young together again by her clucking. When surprised, she utters a well-known danger-signal—a peculiar whine—whereupon the young ones hide under logs and among grass.

"The males never congregate during the breeding-season or after, and I never but once saw two adult males within one-fourth of a mile of each other between April and September. I consider that the drumming is not a call to the female, as they drum nearly or quite as much in the fall as in the spring, and I have heard them drumming every month in the year. I have never seen the least evidence that the Ruffed Grouse is polygamous."

Eggs.—Eight to fourteen is the general number laid; sometimes considerably more are found in a nest. Milky-white and pale buff to pinkish-buff; more or less spotted, but not heavily, with rounded spots and dots of paler reddish-brown.

THE HAZEL-HENS. GENUS TETRASTES.

Tetrastes, Keys. und Blas. Wirbelth. Eur. pp. lxiv. 109, 200 (1840.)

Type, *T. bonasia* (Linn.).

Feet only partially feathered, the lower part being entirely naked; toes naked and pectinate along the sides; tail composed of sixteen feathers, fairly long and bluntly wedge-shaped, the outer feathers being very little shorter than the middle pair. Sexes different. No ruffled frill of fan-shaped feathers on the sides of the neck.

I. THE HAZEL-HEN. TETRASTES BONASIA.

Hazel-Hen, Willoughby, Orn. p. 126, pl. 31 (1676); Lloyd, Game B. Swed. and Norw. p. 112, pl. (1867).

Tetrao bonasia, Linn. S. N. i. p. 275 (1766); Sundev. Svensk Fogl. pl. xxxiii. figs. 4-5 (1856).

Tetrao betulinus, Scop. Ann. i. p. 119, No. 172 (1769).

Bonasia sylvestris, Brehm, Vög. Deutschl. p. 514 (1831); Elliot, Monogr. Tetraon. pl. iv. (1865).

Tetrastes bonasia, Keys. und Blas. Wirbelth. p. 200 (1840);
Ogilvie-Grant, Cat. B. Brit. Mus. xxii. p. 90 (1893).
Bonasa betulina, Dresser, B. Europe, v. p. 193, pl. 486 (1871).

Adult Male.—General colour above greyish or rufous, barred on the head and back with black; feathers of the breast black, margined with white, and sometimes with a white spot in the middle; chin and throat black. Total length, 14 inches; wing, 6·5; tail, 4·9; tarsus, 1·3.

Adult Female.—Differs from the male in having the chin and throat mostly white, and in being rather smaller.

Range.—Europe and North and Central Asia; extending in the west to Scandinavia, in the east to Kamtschatka, Saghalien and Yezo, Japan, and southwards to N. Spain, N. Italy, Transylvania, the Altai Mountains, and N. China.

Habits.—This remarkably handsome Grouse, also known as the "Hazel Grouse" or "Gelinotte" (Hjerpe, Sw. and Norw.), is scarcely larger than the Common Partridge. It inhabits the lower pine-forests, birch-woods, and hazel-copses, being everywhere a local bird and generally confined to the wilder mountainous districts. As an article of food it is very highly esteemed, its white flesh, even after it has been frozen, being most delicious. Von Wright says that "the Finns entertain the very singular notion that, at the creation, this bird was the largest of the feathered tribe; but that year by year it has decreased in size, and will continue to do so until at last it will become so very diminutive as to be able to fly through the eye of a needle; and when that happens the world will come to an end."

Mr. Lloyd, in his "Game Birds and Wild Fowl of Sweden and Norway," gives the following account:—"The flight of the Hazel-Hen is very noisy, but short withal, seldom extending beyond a couple of hundred yards. During both summer and winter it is mostly on the ground, but, when flushed, invariably takes refuge in a tree, rarely on its top, however, as some tell us, but generally about half way up, and amongst the most leafy of the branches. . . . . . .

"The favourite haunts of the Hazel-Hen are hilly and wooded districts. In the open country it is never found, but somewhat varies its ground according to the season of the year. During summer and autumn one often observes these birds in young woods consisting chiefly of deciduous trees; but when the leaves begin to fall, they retire to the great pine-forests, for the reason, as some suppose, that they may be less exposed to the attacks of birds of prey. . . . . . .

"The pairing-season usually commences at the end of March or beginning of April, though the time is somewhat dependent on the state of the weather. The sexes attract each other by a peculiar and almost melancholy cry; that of the male consisting of a long-drawn whistle, followed by a chirp: *ti hīh tītīti-ti*; whilst that of the female is more simple, being often only a single sustained *tih*, vibrating or quivering towards its termination. . . . . .

"The chicks are hatched about midsummer, and in the course of a very few days, and when they are only feathered on the wings and tail, begin to fly."

After describing the various methods employed in Scandinavia for shooting these birds he says:—"The usual way, however, of shooting the Hazel-Hen is without any dog, and solely with the aid of the so-called Hjerp-pipa, or pipe. This implement, which is much less in size that one's finger, and constructed of wood or metal, or it may be the wing-bone of a Black-cock, emits a soft whistling sound, that can be varied according to the call-note of the bird. Such a pipe is readily manufactured. Often, indeed, when we have accidently met with a Hazel-Hen, has my man with his knife alone made one out of a sapling of some pithy tree, and that in the course of a very few minutes. . . . . . .

"The number of Hazel-Hens annually taken in Scandinavia is something enormous. Brunius, in his "Hand Lexicon," published in 1798, calculated that 60,000 were yearly consumed in Stockholm, and 40,000 more in other parts of the country. At the present day that number is, beyond doubt, very greatly exceeded."

Nest.—A small cavity scratched in the ground.

Eggs.—Eight to twelve in number. Pale buff, spotted with brown.

II. THE GREY-BELLIED HAZEL-HEN. TETRASTES GRISEIVENTRIS.

Tetrastes griseiventris, Menzb. Bull. Mosc. lv. pt. i. p. 105, pl. iv. (1880); Ogilvie-Grant, Cat. B. Brit. Mus. xxii. p. 93 (1893).

Adult Male.—Feathers of the breast sandy-grey, with narrow black bars; chin white; the throat black, tipped with dark rufous; the chest reddish-black, barred and tipped with grey. Total length, 14 inches; wing, 6·5; tail, 4·7; tarsus, 1·3.

Adult Female.—Differs in having the chin white; the throat black, tipped with buff; the chest black, irregularly barred with rufous and tipped with grey. Slightly smaller than the male.

Range.—Eastern Russia; Government of Perm.

Nothing is known of the habits of this *perfectly distinct species*, considered by some Russian ornithologists to be merely a variety of the Common Hazel-Hen.

III. SEVERTZOV'S HAZEL-HEN. TETRASTES SEVERTZOVI.

Tetrastes severtzovi, Prjev. Mongolia, ii. p. 130, pl. xviii. (1876): Ogilvie-Grant, Cat. B. Brit. Mus. xxii. p. 93 (1893).

Adult Male.—Feathers of the breast black, barred and tipped with white; outer tail-feathers black, barred with white; chin and throat black. Total length, 13·5 inches; wing, 6·7; tail, 5·3; tarsus, 1·5.

Adult Female.—Differs chiefly in having the chin and throat buff, tipped with black.

Range.—North-eastern Central Asia; Kansu, Koko-nor, and the Hoangho River.

The habits of this species appear to be very similar to those of the Common Hazel-Hen.

THE PARTRIDGES, QUAILS, AND PHEASANTS.
FAMILY PHASIANIDÆ.

Distinguished by having the hind-toe raised above the level of the other toes. The nostrils are never hidden by feathers. The feet (metatarsi) are partially or wholly naked and often armed with spurs. The toes are always naked and never pectinate along the sides, the horny appendages so characteristic of the Bare-toed Grouse being invariably absent.

For convenience' sake this great Family may be divided into the Sub-families *Perdicinæ*, *Phasianinæ*, and *Odontophorinæ*, the first containing the Old World Partridge-like genera, the second the Pheasants and their allies, the Turkeys and Guinea Fowls, and the third the American Partridges and Quails. There appears, however, to be no real line of demarcation between the first two groups, which merge gradually into one another through such forms as the Bamboo Partridges (*Bambusicola*) and the African and Indian Spur-Fowl (*Ptilopachys* and *Galloperdix*). The shape of the wing is perhaps the most important distinguishing mark between the Old World Partridges and Pheasants, and, when taken in connection with the length of the tail, is a useful, if somewhat artificial character. In all the *Perdicinæ*, with but very few exceptions, the first flight-feather is equal to or longer than the tenth, while all the *Phasianinæ*, with the exception of one genus, have the first flight-feather much shorter than the tenth. Unfortunately, the exception among the latter is the important genus *Phasianus*, which has the first flight-feather like that of most Old World Partridges, much longer than the tenth, and, were it not for the long tail, which at once shows it to be a Pheasant, one would certainly feel inclined to place it among the *Perdicinæ*.

By using the combined characters of the shape of the wing and length of the tail, one can artificially separate the two groups, and when a large number of genera have to be dealt with, such divisions, though of no real scientific importance, are at least extremely useful in facilitating the identification of individuals.

The *Perdicinæ* may be characterised as follows :—

OLD-WORLD PARTRIDGES AND QUAILS.
SUB-FAMILY PERDICINÆ.

The cutting-edge of the lower mandible is not serrated. The first flight-feather is equal to or longer than the tenth,* and the tail is shorter, usually much shorter, than the wing. The sides of the head are feathered, with or without a naked space surrounding the eye. The most extreme form of the "Partridge" wing is found in the Snow Partridge (*Lerwa*) and the Quails (*Coturnix* and *Synœcus*), where the first flight-feather is very little shorter than, or sometimes equal to, the second and third, which form the point of the wing.

The extreme form of "Pheasant" wing obtains in the Argus Pheasants (*Argusianus*), where the first flight-feather is the shortest and the tenth the longest.

THE SNOW PARTRIDGES. GENUS LERWA.

Lerwa, Hodgs. Madr. Journ. v. p. 300 (1837); id. Journ. As. Soc. Beng. xxiv. p. 580 (1855).

Type, *L. lerwa* (Hodgs.).

The upper half of the feet (metatarsi) *covered with feathers* as in the Hazel Grouse. Tail about four-sevenths of the length of the wing, rounded, and composed of fourteen feathers. The *first flight-feather equal to the third*, and only slightly shorter

---

* In one or two of the Francolins it is slightly shorter; but the short tail at once distinguishes them as *Perdicinæ*.

than the second and longest. Sexes similar in plumage, but the male is armed with a stout spur on each leg.

Only one species is known.

I. THE SNOW PARTRIDGE. LERWA LERWA.

Perdix lerwa, Hodgs. P. Z. S. 1833, p. 107.
Lerwa nivicola, Hodgs. Madr. Journ. v. p. 301 (1837); Hume and Marshall, Game Birds of India, ii. p. 1, pl. (1879).
Lerwa lerwa, Hartert, Kat. Mus. Senckenb. p. 195 (1891); Ogilvie-Grant, Cat. B. Brit. Mus. xxii. p. 100 (1893).

Adult Male and Female.—Upper-parts black, narrowly barred with whitish; under-parts mostly rich chestnut. Bill and feet red. Total length, 14 inches; wing, 7·5; tail, 4·5; tarsus, 1·7.

Range.—The higher ranges of the Himalayas, and extending northwards to Moupin and Western Sze-chuen, China.

Habits.—This handsome Alpine Partridge is about the size of a Red Grouse, which species it resembles in its strong rapid flight, and in the excellence of its flesh for the table. Mr. Wilson says: "In general haunts and habits, this bird much resembles the Snow Pheasant (*Tetraogallus*), frequenting the same high regions near the snow in summer, and migrating to the same bare hills and rocks in winter. The Pheasant, however, prefers the grassy slopes and softer parts of the hills, the Partridge the more abrupt and rocky portions, where the vegetation is scantier, and more of a mossy than a grassy character. They are also more local, and confined to particular spots, and do not, like the Pheasant, ramble indiscriminately over almost every part of the hill.

"They are generally remarkably tame. When approached, they utter a harsh whistle, and if they keep still, it is often several moments before they can be distinguished, their plumage much resembling and blending with the general colour of much of the ground they frequent. If approached from above, they fly off at once; if from below, they walk away in the opposite direction, calling the whole time, and often cluster

together on the top of some large stone in their way . . .
They seldom fly far, and if followed and put up again, often fly
back to the spot where first found. At times they seem unwilling to get up at all, and several shots may be fired at them
before they take wing."

The Snow Partridge feeds on moss and tender shoots of small plants. It is seldom shot, as those sportsmen who traverse its lonely haunts, which range from 10,000 to 15,000 feet (for it is only met with on lower ground after severe snowstorms), are generally in search of large game such as Tahr and Burrel.

Nest.—"It breeds on the high ridges jutting from the snow at elevations of from 12,000 to 15,000 feet." (*Wilson.*)

Eggs.—" Very large, intermediate in sizes between those of the Chukor (*Caccabis chukor*) and Koklass (*Pucrasia macrolopha*); dull white, freckled *all over* with reddish-brown, like the Koklass, but without blotches of colour." (*Wilson.*)

THE PHEASANT-GROUSE. GENUS TETRAOPHASIS.

Tetraophasis, Elliot, Monogr. Phas. i. pl. xxi. (1871).

Type, *T. obscurus* (Verr.).

The feathers on the feet scarcely extend below the joint; tail wedge-shaped, and rather long, about three-quarters of the length of the wing, and composed of eighteen feathers; first flight-feather about equal to the eighth or ninth, the fourth slightly the longest.

The sexes are similar in plumage, but the male is armed with a stout spur on each leg.

Only two species of these large Grouse-like Pheasants, or rather Partridges, are known from the mountains of Tibet and Western China. Few collections contain examples of these rare birds, and very little is known about their habits.

1. **THE DUSKY PHEASANT-GROUSE. TETRAOPHASIS OBSCURUS.**

Lophophorus obscurus, Verr. N. Arch. Mus. Bull. v. p. 33, pl. vi. (1869).

Tetraophasis obscurus, Elliot, Monogr. Phas. i. pl. xxi. (1871);
  Prjev. in Rowley's Orn. Misc. ii. p. 429 (1877); Ogilvie-
  Grant, Cat. B. Brit. Mus. xxii. p. 102 (1893).

Adult Male and Female.—Above mostly dull olive-brown, barred with buff on the wings; below grey spotted with black, shading into buff on the belly. Distinguished by having the chin, throat, and fore-part of neck dark chestnut. The male measures: Total length, 18·6 inches; wing, 8·3; tail, 6·3; tarsus, 2·2. The female is rather smaller.

Range.—Eastern Tibet, ranging from Moupin to Koko-nor and the mountains of Kansu.

Habits.—Prjevalsky gives the following account of the Dusky Pheasant-Grouse: "We found *T. obscurus* in the same localities in Kansu as the preceding species (the Tibetan Snow Cock), only at a comparatively lower altitude. It was first discovered by Abbé David in Si-chuani (Sze-chuen), and belongs to the middle mountain-ranges, where it principally keeps to the wooded and bush-covered rocks and ravines. Early in spring (about March) they commence pairing, and from that time their voice can be heard daily. It is similar to that of *Crossoptilon auritum* (the Eared Pheasant), being, however, more varied and longer-lasting. The male and female call at the same time, running side by side, with the tail erected and wings dropped. The spreading of the tail is very characteristic; it is like a fan when erected. And this is done also when the bird is surprised or runs in order to avoid danger. When open, the tail is brownish-black, with a distinctly marked white band.

"Like *Crossoptilon auritum*, the present species does not call much, and its voice can be heard only four or five times at certain intervals, but always (or, rather, usually) in the morning at sunrise; and as soon as one pair commences calling, others answer.

"When flushed, *T. obscurus* utters a loud cry, but does not fly fu; and when disturbed in bushes, it always attempts to escape by running. Shooting these birds is extremely difficult."

Nest.—According to the statements of natives, the nest is constructed of grass, on the ground, under thick bushes.

Eggs.—Number unknown; but one lot of four incubated, and another of three fresh, eggs were obtained by a native sportsman for Prjevalsky. The eggs are yellowish-grey or dirty grey, marked with brown spots, which are thickest on the smaller end.

11. SZÉCHENYI'S PHEASANT GROUSE. TETRAOPHASIS SZÉCHENYII.

Tetraophasis széchenyii, Madarász, Zeitschr. ges. Orn. ii. p. 50, pl. ii. (1885); Ogilvie-Grant, Cat. B. Brit. Mus. xxii. p. 103 (1893).

Tetraophasis desgodinsi, Oustalet, Le Nat. 1886, p. 276.

Adult Male and Female.—Distinguished by having the chin, throat, and fore-part of the neck pale fawn-colour. The male measures: Total length, 17·6 inches; wing, 8·9; tail, 6·1; tarsus, 2·2. The female is slightly smaller.

Range.—Mountains of Central Tibet, extending north to the Sok Pass, east to Tà-tsién-loû, and south to Yer-ka-lo, Mekong River.

THE SNOW-COCKS. GENUS TETRAOGALLUS.

Tetraogallus, J. E. Gray, Ill. Ind. Zool. ii. pl. 46 (1833).

Type, *T. himalayensis*, J. E. Gray.

Feathers on the feet scarcely extending below the tarsal joint; tail composed of twenty or twenty-two feathers, rather long, five-eighths of the length of the wing, rounded, the outer pair of feathers being about two inches shorter than the middle pair; first flight-feather about equal to the fifth; an elongate naked patch behind the eye. The feet of the male are provided with a pair of stout spurs.

The six species included in the genus are all large Alpine

birds, the larger forms approaching the Capercailzies (*Tetraogallus*) in size, some males attaining a weight of six and a half pounds.

1. THE TIBETAN SNOW-COCK. TETRAOGALLUS TIBETANUS.

Tetraogallus tibetanus, Gould, P. Z. S. 1853, p. 47; id. B. Asia, vii. pl. 32 (1853); Prjev. in Rowley's Orn. Misc. ii. p. 427 (1877); Hume and Marshall, Game Birds of India, i. p. 275, pl. (1878); Ogilvie-Grant, Cat. B. Brit. Mus. xxii. p. 104 (1893).

Adult Male and Female.—General colour above dark grey and buff; below white, striped with black. Distinguished by having no white on the basal half of the outer quills; chest white, divided from the breast by a grey band. Bill orange-red; feet coral-red. Total length, 20 inches; wing, 10·8; tail, 6·2; tarsus, 2·5.

Range.—Tibet, ranging east to the Sauju Pass, Eastern Turkestan; west to Moupin, north to Kansu and Koko-nor, and south to the Himalayas.

Habits.—In the Himalayas the Tibetan Snow-Cock (known among the Kirghiz as "Utar") appears to be found at elevations of from 15,000 to 19,000 feet, though in the more northern parts of its range, such as Koko-nor, it is met with lower down. The best account of this species is given by Prjevalsky: "Like *C. chukar*, this species is a quick and lively bird; and its voice can almost daily be heard (in north Tibet), at least in spring and summer, in the midst of the wildest and most desolate parts of the mountains. In the middle of the day, however, from about eleven to three o'clock, they do not call, but usually rest; in the morning they begin long before sunrise. . . . .

"In winter they keep in small flocks up to fifteen individuals; and in April, or even earlier, they commence pairing.

"The number of young belonging to a nest varies from five to ten; and we found young ones early in August. They were very small, about the size of a Quail, whilst others were quite as large as their parents. . . . .

"Both parent birds accompany the brood. Whilst the young are small, they crouch on the approach of danger, or try to hide themselves between the loose stones, whilst the old ones keep on running within about twenty paces from the sportsman; but when they are full grown, they try to escape by running, and follow the cock and hen which are leading the whole flock. When much pressed, however, they fly, and do not alight on the ground again until they have crossed a ravine or valley.

"These birds are very wild, and, when alone, the old birds do not allow themselves to be approached within a hundred paces. They hide themselves between stones, and usually spring up and take to flight, or else try to run, which they do so fast that a man cannot catch them.

"We noticed that when they are approached from the bottom of a hill they commence running, but if from the top they at once get up.

"When settling on the ground they shake their tails several times, just as our Willow Grouse do."

Nest and Eggs.—Little or nothing is known. Prjevalsky found a nest containing broken shells, which he believed were evidently of this species. He describes them as " larger than those of the common hen, of a dirty white, shaded with green, and marked on the smaller end with blackish-brown spots."

II. PRINCE HENRY'S SNOW-COCK. TETRAOGALLUS HENRICI.

Tetraogallus henrici, Oustalet, Ann. Sci. Nat (7), xii. pp. 295, 313 (1891); Ogilvie-Grant, Cat. B. Brit. Mus. xxii. p. 106 (1893).

Adult Male and Female.—Said to differ from *T. tibetanus* in having the colour of the throat and chest grey, with only a narrow white band down the middle of the chin and upper half of the throat, and the upper tail-coverts yellowish-grey, not rufous. Total length, 26·4 inches; wing, 11; tail, 7·2; tarsus, 2·2.

I have not examined the typical examples of this species, but I think it very probable that they will prove to be merely younger examples of *T. tibetanus*, which agree closely with the above description. The difference in size is probably due to individual differences in the mode of measuring and to the "make" of the skins, which have perhaps been unnaturally stretched.

III. THE ALTAI SNOW-COCK.   TETRAOGALLUS ALTAICUS.

Perdix altaicus, Gebler, Bull. Sci. Acad. St. Pétersb. i. p. 31 (1837); iv. p. 30 (1840).
Tetraogallus altaicus, Gray, P. Z. S. 1842, p. 105; Gould, B. Asia, vii. pl. 31 (1853); Ogilvie-Grant, Cat. B. Brit. Mus. xxii. p. 110 (1893).

(*Plate IX.*)

Adult Male and Female.—Easily recognised from *T. tibetanus* by having the sides of the neck grey, and the basal part of the outer (primary) flight-feathers white, but there is no white at the base of the secondaries; from the other species it may be distinguished by its white under-parts and the feathers of the sides being uniform white. Bill blackish horn-colour; feet orange-red.

Male: Total length, 23 inches; wing, 10·9; tail, 6·8; tarsus, 2·4. The *female* is slightly smaller.

Range.—Higher chains of the Altai Mountains.

IV. THE HIMALAYAN SNOW-COCK.   TETRAOGALLUS HIMALAYENSIS.

Tetraogallus himalayensis, Gray, P. Z. S. 1842, p. 105; Gould, B. Asia, vii. pl. 30 (1853); Hume and Marshall, Game Birds of India, i. p. 267, pl. (1878); Oates, ed. Hume's Nests and Eggs Ind. B. iii. p. 46 (1890); Ogilvie-Grant, Cat. B. Brit. Mus. xxii. p. 106 (1893).

PLATE IX.

Adult Male and Female.—General colour above grey and buff; throat white, divided by a dark chestnut band from the upper breast, which is white, barred with black; rest of under-parts dark grey; a large chestnut patch on each side of the nape. Basal two-thirds of the outer (primary) flight-feathers white, but the inner (secondary) flight-feathers with only traces of white at the base of the shaft.

Male: Total length, 25·5 inches; wing, 12·6; tail, 7·8; tarsus, 2·8.

Female: Total length, 22·5 inches; wing, 10·8; tail, 6·8; tarsus, 2·5.

Range.—Higher ranges of the Himalayas, extending west to the Hindu Kush and northwards through the Altai Mountains.

Habits.—The following is extracted from Mr. Wilson's account of the Himalayan Snow-Cock, or Snow Pheasant, known as the "Jer-moonal" in the Hills north of Masuri. "It is confined exclusively to the snowy ranges, or the large spurs jutting from them which are elevated above the limits of forest, but is driven by the snows of winter to perform one, and in some places two, annual migrations to the middle regions; in summer they are only seen near the limits of vegetation. In Kunawar they are common at all seasons from Cheenee upwards, but on the Gangetic hills, from June till August, however much a person wanders about on the highest accessible places, but few are met with, and I have no doubt whatever but that nearly all those, which at other seasons frequent this part, retire across the snow into Chinese Tibet to breed. About the beginning of September they are first seen near the tops of the higher grassy ridges, jutting from the snow and the green slopes above, and about the limits of forest. After the first general and severe fall of snow they come down in numbers on to some of the bare exposed hills in the forest regions, and remain there till the end of March. This partial migration is probably made in the night after the fall of snow, as I have invariably found them in their winter quarters early the next morning. It requires a deep fall to drive them down, and in some mild winters, except a few odd birds, they do not come at all. The birds on each respective hill seem to have a particular spot for their winter resort, which they return to, every year the migration is made.

"The Snow Pheasant is gregarious, congregating in packs, sometimes to the number of twenty or thirty, but in general not more than from five to ten; several packs inhabiting the same hill. In summer the few which remain on our side are found in single pairs generally, but across the snow, where the great body migrate, I almost always, even then, found several together. They seldom leave the hill on which they are located, but fly backwards and forwards when disturbed. . . . .

"The Jer-moonal never enters forests or jungle, and avoids spots where the grass is long, or where there is underwood of any kind. It is needless to add that it never perches. During the day, if the weather be fine and warm, they sit on the rocks or rugged parts of the hill without moving much about, except in the morning and evening. When cold and cloudy, and in rainy weather, they are very brisk, and are moving about and feeding all day long. When feeding, they walk slowly uphill, picking up the tender blades of grass and young shoots of plants, occasionally stopping to scratch up a certain bulbous root, of which they seem very fond. If they reach the summit of the hill, after remaining stationary some time, they fly off to another quarter, alighting some distance down, and again picking their way upwards. When walking, they erect their tails, have a rather ungainly gait, and at a little distance have something the appearance of a large Grey Goose. . . .

"The Jer-moonal is not remarkably wild or shy. When approached from below, on a person getting within eighty or a hundred yards, they move slowly uphill or slanting across, often turning to look back, and do not go very far unless followed. If approached from above, they fly off at once, without walking many yards from the spot. They seldom, in any situation, walk far downhill, and never run except for a few yards when about to take wing."

Nest.—A hole scratched in the ground, under the shelter of a stone, rock, or bush, at elevations of from 12,000 to 17,500 feet.

Eggs.—Five in number generally, but said to be as many as nine and even twelve. In shape a long perfect oval; shell minutely pitted with pores; olive or brownish stone-colour, with numerous spots and dots, and sometimes small blotches of reddish- or purplish-brown. Average measurements, 2·72 by 1·85 inches.

V. THE CASPIAN SNOW-COCK. TETRAOGALLUS CASPIUS.

Tetrao caspius, Gmel. Reise, iv. p. 67, pl. 10 (1784).
Tetraogallus caspius, Gould, B. Asia, vii. pl. 29 (1853); Sclat. in Wolf's Zool. Sketches (1), pl. 40 (1861); Ogilvie-Grant, Cat. B. Brit. Mus. xxii. p. 108 (1893).
Megaloperdix raddei, Bolle and Brehm, J. f. O. 1873, p. 4.
Tetraogallus challayei, Oustal. Bull. Soc. Philom. 1875, p. 54.
Tetraogallus tauricus, Dresser, P. Z. S. 1876, p. 675.

Adult Male.—Like *T. himalayensis,* but paler in its general colour, and easily distinguished by the grey chest, the absence of chestnut on the sides of the nape and head, and by having the basal part of the inner (secondary) flight-feathers white. Total length, 24 inches; wing, 11·5; tail, 7·5; tarsus, 2·6.

Adult Female.—Differs in having the grey feathers of the chest mottled with buff. Total length, 23 inches; wing, 11; tail, 6·9; tarsus, 2·3.

Range.—Mountains of Asia Minor, ranging west of the Gök Mountains, east to Transcaspia, north to the Caucasus, and south to the higher ranges near Shiraz, S. Persia.*

* Possibly the bird from S. Persia may be different. Mr. Hume thinks it *may* prove to be *T. himalayensis.*

VI. THE CAUCASIAN SNOW-COCK. TETRAOGALLUS CAUCASICUS.

Tetrao caucasica, Pall. Zoogr. Ross.-As. ii. pp. 76, 87 (note), pl. (1811).
Chourtka alpina, Motschoulski, Bull. Soc. Mosc. No. i. p. 95, pls. viii. viii. *bis* and ix. (1839.)
Tetraogallus caucasicus, March, Rev. Zool. 1877, p. 354, pl. 133; Ogilvie-Grant, Cat. B. Brit. Mus. xxii. p. 109 (1893).
Megaloperdix caucasica, Radde, Orn. Caucas. p. 335, pl. xxi. figs. 1 and 2 (1884).

Adult Male and Female.—Resemble *T. caspius* in general plumage, but the back of the head and nape are rust-red, and there is a dull chocolate band down each side of the throat; the whole upper back is barred and mottled with black and buff, and the chest is blackish-grey, irregularly barred and mottled with buff. Total length, 21 inches; wing, 10·8; tail, 6·8; tarsus, 2·4.

Range.—Higher ranges of the Caucasus Mountains.

THE RED-LEGGED PARTRIDGES. GENUS CACCABIS.

Caccabis, Kaup, Natürl. Syst. p. 183 (1829).

Type, *C. saxatilis* (Wolf and Meyer).

The feathers on the feet scarcely extending below the tarsal joint; tail composed of fourteen feathers, somewhat rounded, and five-eighths of the length of the wing; first flight-feather about equal to the sixth, third slightly the longest; throat covered with feathers; sub-terminal part of the outer webs of the outer (primary) flight-feathers buff; sides and flanks *transversely barred*, in marked contrast to the rest of the plumage of the under-parts. *Sexes similar.* Male provided with a pair, or sometimes more, of stout blunt spurs.

1. THE ROCK RED-LEGGED PARTRIDGE. CACCABIS SAXATILIS.

Perdix saxatilis, Wolf and Meyer, Hist. Nat. Ois. Allem. p. 87, pl. 48 (1805); Gould, B. Europe, iv. pl. 261, fig. 2 (1837); Gigl. Iconogr. Av. Ital. pl. 252 (1881).

Caccabis saxatilis, Dresser, B. Europe, vii. p. 93, pl. 470 (1875);
Ogilvie-Grant, Cat. B. Brit. Mus. xxii. p. 111. (1893).

Adult Male and Female.—Above greyish olive brown; breast grey; belly pale rufous-buff; outer tail-feathers dark chestnut; top of the head dull vinaceous-grey; white throat and fore-neck surrounded by a black band; feathers of the chest uniform, not margined with black on the sides; *lores black;* ear-coverts black, mixed with buff.

Male : Total length, 15 inches; wing, 6·6; tail, 3·6; tarsus, 1·8.

Female : Total length, 13·6; wing, 6·2; tail, 3·4; tarsus, 1·7.

Range.—Mountains of Southern Europe. Eastern Pyrenees, Alps, Carpathians, Apennines, and Balkans; also Sicily. It still remains uncertain whether it is this species or the closely allied form, *C. chukar*, which is found in the mainland of Greece; it is certainly the latter which is met with in the Grecian Archipelago, but so far I have been unable to obtain examples of the mainland bird.

Hybrids.—Crosses have been described between this species and the Barbary Red-legged Partridge (*C. petrosa*); and also with the Common Red-legged Partridge (*C. rufa*).

Habits.—This species inhabits the desolate stony hillsides, and its mode of life and habits are very similar to those of its eastern ally, the Chukar, which are fully described below.

Professor Victor Fatio records a curious variety of this species, with a black head. For this bird, of which he has seen three examples from Switzerland, he proposes the name of *C. saxatilis*, var. *melanocephalus* (*nec* Rüpp.).

SUB-SP. *a.* THE CHUKAR RED-LEGGED PARTRIDGE. CACCABIS CHUKAR.

Perdix chukar, J. E. Gray, Ill. Ind. Zool. i. pl. 54 (1830-32);
Gould, Cent. B. Himal. pl. 71 (1832).

Caccabis chukar, G. R. Gray, List of B. pt. iii. Gall. p. 36 (1844); Ogilvie-Grant, Cat. B. Brit. Mus. xxii. p. 113 (1893).

Caccabis chukor, Hume and Marshall, Game Birds of India, ii. p. 34, pl. (1879); Oates, ed. Hume's Nests and Eggs Ind. B. iii. p. 431 (1890).

Adult Male and Female.—Closely resemble the western form, *C. saxatilis*, but always differ in having the lores, or space immediately behind and below the nostril scale, white or whitish-buff instead of black,* and the ear-coverts chestnut.

Male: Total length, 14·6 inches; wing, 6·7; tail, 3·9; tarsus, 1·9.

Female: Total length, 13·4 inches; wing, 6·5; tail, 3·8; tarsus, 1·8.

Range.—Asia, extending in the west to the Ionian Islands [and possibly the mainland of Greece], in the east to China, in the north to Mongolia and Turkestan, and in the south to the Persian Gulf and possibly to Arabia. Island of St. Helena [introduced].

This bird varies immensely in size and colour in different localities, but all the various forms pass imperceptibly into one another and must be regarded as mere climatic varieties of the same sub-species. The lightest coloured birds in all the large series I have examined come from the arid neighbourhood of Bushire at the head of the Persian Gulf. Somewhat darker forms occur at Bagdad and Shiraz, in Afghanistan, Sind, Ladak, and other localities where the physical surroundings are somewhat similar in character, while the darkest and most richly-coloured examples are those from the Ionian Islands, Cyprus, Asia Minor, and the outer Himalayas, where vegetation is more plentiful. In birds from North China, the upper-parts of the body have a more reddish tint, but specimens from northern

---

* It must however be noted that some specimens of *C. chukar* have a very small spot of black feathers below the nasal opening, thus approaching *C. saxatilis*.

Afghanistan and several other localities approach them closely in colour.

Mr. Hume says: "The Chukor may be found in different localities from sea-level, as in Southern Sind and Beluchistan, to an elevation of at least 16,000 feet, as in Ladak and Tibet

"It will be found in comparatively well wooded, watered, and cultivated hills, as throughout the lower, southern, or outer ranges of the Himalayas; in absolute deserts, like those of Ladak and the Karakoram plateaux; or in utterly barren rocky ranges, like those of the Mekran and Arabian coasts, where the abomination of desolation seems to reign enshrined.

"In one place it faces a noon-day temperature of 150° Fahr.; in another, braves a cold, about daybreak, little above zero; here it thrives where the annual rainfall exceeds 100 inches, and there flourishes where it is practically *nil*. But all these differences in physical environment affect appreciably the size and colour of the species; and hence the numerous races which, under a variety of names (*rupicola, altaica, sinaica, pallescens, pubescens, arenarius, pallidus, &c.*), have been at one time or another elevated to the rank of species. . . . .

"The Chukor is a very noisy bird, repeating constantly in a sharp, clear tone, that may be heard for a mile or more through the pure mountain air, his own well-applied trivial name. Like other Game-Birds, they call most in the mornings and evenings; but even when undisturbed, they may be heard calling to each other at all hours of the day; and very soon after a covey has been dispersed, each individual member may be heard proclaiming his own, and anxiously enquiring after all his fellows' whereabouts. The tone varies. First he says, 'I'm here, I'm here'; then he asks, 'Who's dead? Who's dead?'; and when he is informed of the untimely decease of his pet brother and favourite sister, or perhaps his eldest son and heir, he responds, 'Oh lor! Oh lor!' in quite a mournful tone."

The following account of its habits are given by Mr. Wilson.

"In our part of the hills the Chukor is most numerous in the higher inhabited districts, but is found scattered over all the lower and middle ranges. In summer they spread themselves over the grassy hills to breed, and about the middle of September begin to assemble in and around the cultivated fields near the villages, gleaning at first in the grain-fields which have been reaped, and afterwards, during winter, in those which have been sown with wheat and barley for the ensuing season, preferring the wheat. A few straggling parties remain on the hillsides, where they breed, as also in summer many remain to perform the business of incubation in the fields. In autumn and winter they keep in loose scattered flocks, where numerous, sometimes to the number of forty or fifty, or even a hundred. In summer, though not entirely separated, they are seldom in large flocks, and a single pair is often met with. They are partial to dry, stony spots, never go into forest, and in the lower hills seem to prefer the grassy hillsides to the cultivated fields. This may probably be owing to their comparatively fewer numbers, as I have observed that many others of the feathered race are much shyer and more suspicious of Man when rare, than those of the same species in places where more numerous. . . . . .

"The Chukor feeds on grain, roots, seeds, and berries; when caught young it soon becomes tame, and will associate readily with domestic poultry.

"From the beginning of October, Chukor-shooting, from the frequency and variety of the shots, and the small amount of fatigue attending it, is, to one partial to such sport, perhaps the most pleasant of anything of the kind in the hills. About some of the higher villages, ten or a dozen brace may be bagged in a few hours. Dogs may be used or not, at the discretion of the sportsman; they are not necessary, and if at all wild, are more in the way than otherwise."

Nest.—Composed of leaves and fibres, placed in a depression in the ground, and generally sheltered by a tuft of grass or low bush; may be met with in different localities from the sea-level up to an elevation of 16,000 feet.

Eggs.—Generally seven to twelve in number, sometimes more; somewhat sharply pointed. The ground-colour varies from yellowish-white to brownish-cream, thickly speckled and spotted with purplish or reddish-brown. Average measurements, 1·68 by 1·25 inch.

11. PRJEVALSKY'S RED-LEGGED PARTRIDGE. CACCABIS MAGNA.

Caccabis magna, Prjev. in Rowley's Orn. Misc. ii. p. 426 (1877); Ogilvie-Grant, Cat. B. Brit. Mus. xxii. p. 120 (1893).

Adult Male and Female.—Differ from *C. saxatilis* in their paler colour and larger size, and in having the collar round the base of the neck *double, inside blackish or black, on the outside reddish-brown*. Total length, 15 inches; wing, 7·6; tail, 4·1; tarsus, 1·7.

Range.—Mountains of South Koko-nor, Northern Tibet, and the Tsaidam Plains.

Habits.—This remarkable and perfectly distinct "Red-leg" was first obtained by the great Russian traveller, Prjevalsky, who makes the following remarks on its habits. He says:—"We first obtained this bird in the most desolate parts of South Koko-nor Mountains; and later on we met with it also in Northern Tibet and the Tsaidam Plains. In its habits it does not differ from *C. chukar*, and keeps usually in small companies (probably families) on the rocky mountains and in the neighbourhood.

"When taking wing it utters a peculiar hollow note, something like '*cuta-cuta*,' which we never noticed in *C. chukar*, and the present species seems to be more silent than the preceding one."

III. THE COMMON RED-LEGGED PARTRIDGE. CACCABIS RUFA.

Tetrao rufus, Linn. S. N. i. p. 276 (1766).
Perdix rubra, Gould, B. Europe, iv. pl. 260 (1837).
Caccabis rufa, Dresser, B. Europe, vii. p. 103, pl. 471, fig. 1 (1875); Ogilvie-Grant, Cat. B. Brit. Mus. xxii. p. 118 (1893).

(*Plate X.*)

Adult Male and Female.—Resemble *C. saxatilis* in general appearance, but are darker and more richly coloured. In addition to the black band which circumscribes the throat, the feathers of the chest are *widely margined on the sides with black*, those of the sides and back of the neck more narrowly; the belly is bright rufous-buff and the outer tail-feathers are dark chestnut.

Male: Total length, 13·6 inches; wing, 6·2; tail, 3·7; tarsus, 1·7.

Female: Total length, 13 inches; wing, 6; tail, 3·6; tarsus, 1·6.

Range.—South-western Europe; ranging in the north to Belgium and Switzerland; in the south to Madeira, the Azores, and Gran Canary; in the west to North and Central Italy. It is also found in Elba, Corsica, the Balearic Islands, and in Great Britain [introduced].

In Spain a somewhat darker and more richly coloured climatic variety of *C. rufa* is met with, which has been named *Caccabis rufa hispanica* by Prof. Seoane. This form is figured in the accompanying plate.

Habits.—This remarkably handsome species was first introduced into the south-eastern counties of Great Britain about a century ago. Like the rest of its allies, it is an inveterate runner, and generally prefers to escape from approaching danger on foot, which it does with great rapidity, seldom taking to flight unless hard pressed or suddenly disturbed. When once on the wing, however, the flight is rapid and straight, and

SPANISH RED-LEGGED PARTRIDGE.

for this reason these birds afford capital sport when driven; but if shot over dogs or walked up in cover their cursorial habits are alike detestable to Man and Dog, for the Red-legs not only seldom rise themselves till they are at the other end of the field and probably far out of shot, but disturb and put up any coveys of Grey Partridges they may chance to pass on their course. They are very partial to hedgerows or the edges of plantations and long grass or rushes, and when flushed, occasionally perch on a neighbouring tree, which the Grey Partridge, so far as we are aware, never does. In the pairing-season the Red-legs are very pugnacious, fighting fiercely not only with the males of their own kind, but also with those of the Grey Partridge, which, being much smaller birds, are in most cases driven from the field. Eggs of the latter species, as well as those of the Common Pheasant, are sometimes found in the nests of *C. rufa*, and are doubtless laid there by the females instead of in their own nest, an irregular habit by no means rare among Game-Birds.

Nest.—A hollow scratched in the ground under the shelter of a hedge, tall grass, or growing crops.

Eggs.—Ten to eighteen in number, and sometimes more. Like those of *C. saxatilis* and *C. chukar*, pale stone-colour or buff, more or less thickly dotted and spotted, and sometimes blotched with dark reddish-brown. Average measurements, 1·55 by 1·2 inch.

IV. BARBARY RED-LEGGED PARTRIDGE. CACCABIS PETROSA.

Tetrao petrosus, Gm. S. N. i. pt. ii. p. 758 (1788).

Perdix petrosa, Lath. Ind. Orn. ii. p. 648 (1790); Gould, B. Europe, iv. pl. 261, fig. 1 (1837).

Perdix barbara, Bonn. Tabl. Encycl. Méth. i. p. 208, pl. 94, fig. 2 (1791).

Caccabis petrosa, Dresser, B. Europe, vii. p. 111, pl. 471, fig. 2 (1875); Ogilvie-Grant, Cat. B. Brit. Mus. xxii. p. 120 (1893).

Adult Male and Female.—Easily recognised from the species already mentioned by having the top of the head dark chestnut, a wide chestnut collar spotted with white bordering the sides and front of the neck; and the outer scapulars bordered with rufous-chestnut instead of vinaceous or grey.

Male: Total length, 12·5 inches; wing, 6·5; tail, 4·1; tarsus, 1·8.

Female: Smaller; wing, 6·1.

Range.—This extremely handsome species has a comparatively limited range, being found in North-west Africa, Sardinia, near Gibraltar, and in some of the islands of the Canary group. In both the last-named localities it has doubtless been introduced. Specimens have been obtained in Malta, but whether such examples are escaped cage-birds or accidental migrants is uncertain.

Eggs.—Similar to those of *C. rufa*, but the ground-colour is usually more rufous.

V. THE BLACK-HEADED RED-LEGGED PARTRIDGE. CACCABIS MELANOCEPHALA.

Perdix melanocephala, Rüpp. Neue Wirb. Vög. p. 11, pl. v. (1835).

Caccabis melanocephala, Gray, Gen. B. iii. p. 508 (1846); Yerbury, Ibis, 1886, p. 19; Barnes, Ibis, 1893, p. 166; Ogilvie-Grant, Cat. B. Brit. Mus. xxii. p. 122 (1893).

Adult Male and Female.—General colour slaty-grey, shading into buff on the under-parts. Top of the head *black;* a wide black band surrounding the throat and continued down the middle of the neck; *outer tail-feathers grey.*

Male: Total length, 16·6 inches; wing, 7·7; tail, 5·7; tarsus, 22·5.

The *female* is somewhat smaller; wing, 7·2.

Range.—South-west Arabia, Jeddah, and Mecca, to Aden.

Habits.—Very little is known about the habits of this fine

Red-leg, considerably the largest species of the group. Lieut. Barnes, writing from Aden, says that it "is common in the ravines at the base of the hills, some distance inland. They also frequent the clayey cliffs along the river banks, especially near pools of water, the river-bed being generally dry."

THE SEESEE PARTRIDGES. GENUS AMMOPERDIX.

Ammoperdix, Gould, B. Asia, vii. pl. 1 (part iii.; 1851).

Type, *A. bonhami* (Fraser).

The feathers on the feet scarcely extending below the tarsal joint; tail composed of *twelve feathers*, somewhat rounded, but the feathers sub-equal, and about half the length of the wing; first flight-feather about equal to the sixth, and not much shorter than the third and longest; bill yellowish; no space behind the eye or on the cheeks; throat covered with feathers; flanks of the male *longitudinally barred*, in marked contrast to the rest of the plumage of the under-parts. *Sexes different.* No trace of spurs in either sex.

Only two rather small species are known.

1. BONHAM'S SEESEE PARTRIDGE. AMMOPERDIX BONHAMI.

Perdix bonhami, Fraser, P. Z. S. 1843, p. 70.
Caccabis bonhami, Fraser, Zool. Typ. pt. 3, pl. 61 (1849).
Perdix griseogularis, Brandt, Bull. Ac. St. Pétersb. i. p. 365 (1843).
Ammoperdix bonhami, Gould, B. Asia, vii. pl. 1 (1851); Hume and Marshall, Game Birds of India, ii. p. 45, pl. (1879); Dresser, B. Europe, vii. p. 117, pl. 472 (1880); Oates, ed. Hume's Nests and Eggs Ind. B. iii. p. 433 (1890); Ogilvie-Grant, Cat. B. Brit. Mus. xxii. p. 123 (1893).

Adult Male.—General colour isabelline; a black band across the forehead, continued backwards in eyebrow stripes; chin whitish; cheeks, throat, and front of neck grey; flank-feathers vinaceous and chestnut, margined on either side with black. Total length, 9·5 inches; wing, 5·2; tail, 2·4; tarsus, 1·2.

Adult Female.—Differs chiefly from the male in having the black and white markings on the head and the barring on the flanks absent. Total length, 9·5 inches; wing, 4·9; tail, 2·3; tarsus, 1·15.

Range.—South-western Asia, extending westwards to the Euphrates Valley, eastwards to North-west India, in the north to Transcaspia, and south to Aden.

Habits.—This handsome little Partridge is met with at elevations ranging from sea-level to 6,000 or 7,000 feet. They are particularly common in the Salt Range, and Mr. Hume gives us the following account of their habits: "They are eminently birds of bare broken ground; on grassy slopes they may indeed be found, for they feed much on grass-seeds, but they eschew utterly forests or thickly-wooded tracts, and even where there is much scrub about they are less common—the barer and more desolate the ravines and gorges, the more thoroughly do they seem at home.

"They are active, bustling little birds, scratching about a great deal in the earth, dusting themselves freely in the sand, basking in the sun, resting in little hollows they have worked out for themselves, and generally reproducing in many ways the manners of the Domestic Fowl.

"Their call, continually heard in the spring, is a clear double note, "Soo-see, soo-see," and they have also, whilst feeding and when surprised, a whistled chirp, uttered very softly when at their ease, but sounding more harshly when they are alarmed.

"Their food is, I think, chiefly, if not exclusively, grain, seeds, and herbage of different kinds. I have examined many, but have lost my notes in regard to them, and I cannot now remember whether they are or are not also insectivorous. My impression is that they are not.

"Although they are pretty shooting, they never afford much sport; they run a great deal, and over ground across which it is difficult to follow them; it is often difficult to flush them, and when flushed they constantly rise so little, and dart so directly downhill, that they are lost sight of before it is possible to fire."

Nest.—Very slight; a few blades of dry grass laid in a depression in the ground under a bush or a ledge of rock or among stones.

Eggs.—Vary in number from eight to fourteen, and sometimes more are laid; lengthened ovals, generally somewhat pointed towards one end. The colour varies from nearly white to stone-cream; shell somewhat glossed and minutely pitted. Average measurements, 1·42 by 1·02 inch.

II. HEY'S SEESEE PARTRIDGE. AMMOPERDIX HEYI.

Perdix heyi, Temm. Pl. Col. v. pls. 37, 38 [Nos. 328, 329] (1825).

Caccabis heyii, Gray, List Gall. Brit. Mus. p. 37 (1844); Wyatt, Mamm. and Avif. Sinai, pl. xix. (1873).

Ammoperdix heyi, Gould, B. Asia, vii. pl. 2 (1851); Ogilvie-Grant, Cat. B. Brit. Mus. xxii. p. 125 (1893).

Adult Male.—Differs chiefly from the male of *A. bonhami* in having the general colour of the upper-parts much paler; no black band across the forehead or above the eyes; the chin and middle of throat chestnut. Total length, 9·5 inches; wing, 5; tail, 2·5; tarsus, 1·25.

Adult Female.—Resembles the female of *A. bonhami*.

Range.—Both sides of the Red Sea, extending north to the Dead Sea, westwards to Egypt and Nubia, about as far south as 20° N. latitude, and eastwards to Muscat, Persian Gulf.

THE FRANCOLINS. GENUS FRANCOLINUS.

Francolinus, Steph. in Shaw's Gen. Zool. xi. pt. ii. p. 316 (1819).

Type, *F. francolinus* (Linn.).

The feathers of the feet scarcely extend below the tarsal

joint; tail composed of fourteen feathers, half the length of the wing, or rather more; first flight-feather varying in length between the seventh and tenth*; the fourth to the sixth forming the angle of the wing; throat covered with feathers; plumage of the flanks not barred, or, if barred, not contrasting with the rest of the under-parts. Sexes usually similar, or nearly similar, in plumage, but in a few species extremely different. Feet without spurs, or with one or more pairs.†

Although certain of the large number of species forming the various groups of this genus differ considerably from one another in several important points, and have in consequence received a variety of generic or sub-generic names, I have so far found it impossible to divide the genus *Francolinus* into minor sections, the less highly characterised species forming intermediate links which prevent any of the proposed divisions from being satisfactorily characterised.

To assist in the identification of the forty-four species comprising this great genus, the various groups of allied forms have been divided under several headings, characterised by prominent differences in the marking of the plumage.

A. A well-defined row of rufous or buff spots on *both* webs of the primary flight-feathers (species 1 to 3, pp. 103-107).

B. No well-defined row of rufous or buff spots on both webs of the primary flight-feathers; feathers of the back and scapulars with white or buff shaft-stripes down the middle.

    a. Breast and flanks whitish buff, with *uniform transverse bars of black* (species 5 to 8, pp. 108-112).
    b. Breast and flanks *not* whitish buff, barred with black (species 9 to 26, pp. 112-122).

---

* In *Francolinus squamatus* and *F. schuetti* the first flight-feather is slightly shorter than the tenth, so that the shape of the wing is somewhat Pheasant-like, but the shortness of the tail at once distinguishes these birds as *Perdicinæ*.

† In many of the species, the females have no spurs, but it is not uncommon to find a blunt pair developed in old birds.

COMMON FRANCOLIN

C. *No* well-defined row of rufous or buff spots on both webs of the primary flight-feathers; feathers of the back and scapulars *devoid* of white or buff shaft-stripes down the middle (p. 124).

    c. Inner webs of the primary flight-feathers either mostly pale buff or brown, largely barred and mottled with chestnut or buff (species 28 to 34, pp. 125-128).

    d. Inner webs of the primary flight-feathers uniform dark brown, sometimes slightly dotted with buff towards the marginal extremity (species 35 to 44, pp. 129-135).

The Francolins or Spur-legged Partridges vary much in size, some being not much larger than Quails, others rather larger than the Red-legged Partridges.

A. The three following Asiatic species are characterised by having *a well-defined row of rufous or buff spots on both webs of the primary flight-feathers.*

1. THE COMMON FRANCOLIN. FRANCOLINUS FRANCOLINUS.

Tetrao francolinus, Linn. S. N. i. p. 275 (1766).
Perdix francolinus, Vieill. Faun. Franç. p. 254, pl. 110, fig. 2, and pl. iii. fig. 1 (1828).
Francolinus vulgaris, Steph. in Shaw's Gen. Zool. xi. p. 319 (1819); Gould, B. Europe, iv. p. 259, pl. (1837); Dresser, B. Europe, vii. p. 123, pl. 473 (1876); Hume and Marshall, Game Birds of India, ii. p. 9, pl. (1879); Oates, ed. Hume's Nests and Eggs Ind. B. iii. p. 428 (1890); Ogilvie Grant, Cat. B. Brit. Mus. xxii. p. 132 (1893).
Francolinus tristriatus, *F. henrici*, and *F. asiæ*, Bonap. C. R. xlii. p. 882 (1856).

(Plate XI.)

Adult Male.—General colour of the under-parts black, spotted with white on the sides; upper-back black, spotted with white; lower-back barred with white; *a wide dark chestnut nuchal collar*; a white patch on the hinder-part of the cheek; *rest of head and throat black.* A pair of small wart-shaped spurs.

Total length, 13 inches; wing, 6; tail, 3·5; tarsus, 1·7 (Indian specimens). Examples from Cyprus, Asia Minor, Persia, &c., have larger dimensions; wing, 6·9.

Adult Female.—Differs chiefly from the male in having the colour of the upper-parts browner; the sides of the face buff, dotted with black; the throat white; the chestnut collar confined to the nape, and the under-parts whitish-buff, more or less strongly-marked with V-shaped black bars. Total length, 12·6 inches; wing, 5·9; tail, 3·5; tarsus, 1·6 (Indian specimens). Examples from Cyprus, &c., larger; wing, 6·6.

Range.—From Cyprus, Palestine, and Asia Minor, through Persia eastwards to Northern and Central India, to Assam and southward to Manipur. Formerly found in Sicily, but now apparently extinct. It also, no doubt, occurred in Sardinia, Spain, and on the north-west coast of Africa, but has long since been exterminated.

Habits.—This handsome species, also known as the Black Partridge, or *Kala titur* among Indian sportsmen, is still numerous in many parts of Upper India, and affords most excellent shooting, being either bagged from elephants with a close line of beaters, or shot over dogs.

Mr. Hume publishes the following notes on this species, sent to him by Mr. O. Greig: "The Francolin is not a prolific breeder. I hardly ever remember to have seen more than three young ones in a brood. Probably, being a ground bird, the young are killed by stoats, jackals, and other vermin, and the mother is not of sufficient size to defend them. It seems to have a second brood sometimes.

"It remains entirely on the ground, as a rule, except the cock when calling, when he will at times get on to a stump or ant-hill; but up the Touse Valley, and in the Rama Serai, in Native Garhwál, I have seen them high up in chir-trees (*Pinus longifolia*).

"From its breeding so slowly it is easily shot off, and I have known a place almost cleared in one season. The Western

Dún has been served in that way. Formerly twenty-five brace could be bagged there, but now, if a man flushes five brace in a day, he has done well."

"All sportsmen who like Black Partridge shooting should kill all vermin they see about its haunts.

"This bird gets tame readily, and, even when caught full grown, will feed on the day it is caught. It affords some of the finest sport of all small game, and with steady dogs one may have grand shooting. It may be found in all crops, but especially in cotton-fields freshly sown, wheat, rice, and mustard, and in wild hemp. It runs a good deal at times, but will lay like a stone if headed; it is never found far from grass-jungles.

"Some hens have spurs of the same size and shape as the cocks.

"It is kept tame by the natives, and used for the capture of wild ones in the breeding-season. The mode of using it is to put it in a cage out near wild ones in the pairing-season, and to set snares round the cage. The tame ones then call up the wild ones; but only cocks are caught in this way, and the tame one must be a young one reared by hand, as, if caught when old, it will not call.

"Netting is largely used to capture this bird, and on one occasion I wanted some birds to stock a bit of forest, and a man caught two score of birds in a very short time.

"I never heard of this bird being used for fighting; it is merely kept as a call-bird or as a pet."

Nest.—Always well hidden; often slight, sometimes more substantial, and composed of grass, roots, and dry bamboo, &c.; placed in a hollow in the ground, at elevations varying from nearly sea-level to 6,000 feet.

Eggs.—Six to ten in number, bluntly pointed at the smaller end, and varying in colour from uniform greenish stone-colour to rich brownish-buff. Average measurements, 1·56 by 1·28 inch.

II. THE PAINTED FRANCOLIN. FRANCOLINUS PICTUS.

Perdix picta, Jard. and Selb. Ill. Orn. pl. 50.
Perdix hepburnii, J. E. Gray, Ill. Ind. Zool. i. pl. 55, fig. 1 (1830-32).
Francolinus pictus, Hume and Marshall, Game Birds of India, ii. p. 19, pl. (1879); Legge, B. Ceylon, iii. p. 744 (1880); Oates, ed. Hume's Nests and Eggs Ind. B. iii. p. 430 (1890); Ogilvie-Grant, Cat. B. Brit. Mus. xxii. p. 138 (1893).

Adult Male.—Differs from the male of *F. francolinus* in having *no trace* of a chestnut collar; the forehead and sides of the head rust-red; the throat paler rufous, spotted with black; the *scapulars black, edged with buff;* the under-parts black, covered with round white spots. Spurs entirely wanting.

Adult Female.—Very similar to the male, but the throat is whitish and not spotted with black.

Total length, 11·6 inches; wing. 5·3; tail 2·5; tarsus, 1·7.

Range.—Western and Central India, extending in the west to North Guzerat, northwards to Hamirpur, and south as far as Coimbatore on the east and Masulipatam on the west. It is also found in Ceylon.

Habits.—This Painted Francolin is very locally distributed over its range, and is far more arboreal in its habits than *F. francolinus*, which rarely perches. It may often be met with roosting on bushes and trees. Its favourite haunts are dry fields studded with trees, the higher uplands covered with scrub-jungle, or broken hilly ground, and it avoids the damper lower-lying country where the Common Francolin is ordinarily met with.

Hybrid.—This species is known to cross with *F. francolinus*, and Colonel E. A. Butler shot six or seven such hybrids near Deesa, a locality where the ranges of the two species meet.

For a figure of one of these hybrids see Hume and Marshall, Game B. India, ii. p. 27, pl. fig. 2 (1879).

Nest and Eggs.—Very similar to those of *F. francolinus*, but larger and generally less like a peg-top in shape. Average measurements, 1·4 by 1·18 inch.

III. THE CHINESE FRANCOLIN.   FRANCOLINUS CHINENSIS.

Tetrao chinensis, Osbeck, Voy. en Chine, ii. p. 326 (1771).
Tetrao madagascariensis and *T. pintadeanus*, Scop. Del. Flor. et Faun. Insubr. pt. ii. p. 93 (1786).
Tetrao perlatus, Gmel. S. N. i. pt. ii. pp. 756, 758 (1788).
Perdix (*Francolinus*) *maculatus*, Gray, Fasc. B. China, pl. 7 (1871).
Francolinus phayrei, Blyth, J. As. Soc. Beng. xii. p. 1011 (1843).
Francolinus chinensis, Hume and Marshall, Game Birds of India, ii. p. 27, pl. (1879); Oates, ed. Hume's Nests and Eggs Ind. B. iii. p. 431 (1890); Ogilvie-Grant, Cat. B. Brit. Mus. xxii. p. 137 (1893).

Adult Male.—Like *F. pictus*, this species has *no chestnut collar*, but differs in having the scapulars chestnut, or chestnut and black, with rounded spots of white or buff. A black band crosses the forehead and is continued behind the eye, and a second, starting from the angle of the gape, crosses the cheek; rest of the sides of the head and throat white. Feet armed with a pair of sharp spurs.

Adult Female.—Differs from the male in having the sides of the head washed with rufous, the scapulars *black, margined with brownish*, and spotted and barred with buff, and the under-parts buff, barred with black.

Total length, 12·6 inches; wing, 5·5; tail, 3; tarsus, 1·6.

Range.—Indo-Chinese countries, Burma, Siam, Cochin China, Hainan, and Southern China. It is not found in Tenasserim. It was introduced more than a century ago into Réunion and Mauritius. ? Madagascar.

Habits.—Very similar to those of *F. pictus*, but it seldom visits the open country, preferring near Thayetmyo, where it is specially numerous, the "gravel hills with bamboo-jungle, intermingled with abandoned clearings, in the dense vegetation of which it loves to conceal itself" (*Oates*).

B. All the following species (Nos. 4 to 25 inclusive) are characterised by having *no well-defined row of buff spots on the inner and outer webs of the primary flight-feathers, but the feathers of the back and scapulars have white or buff shaft-stripes down the middle.* The following species *only* has the throat black; *in all the rest it is differently coloured.*

IV. LATHAM'S FRANCOLIN. FRANCOLINUS LATHAMI.

Francolinus lathami, Hartl. J. f. O. 1854, p. 210; Ogilvie-Grant, Cat. B. Brit. Mus. xxii. p. 139 (1893).
Francolinus peli, Temm. Bijdr. tot de Dierk. I. p. 50, pl. (1854).

Adult Male.—General colour above olive-brown; *throat and fore-neck black;* breast black, each feather with a white heart-shaped spot. Total length, 10 inches; wing, 5·6; tail, 2·7; tarsus, 1·7.

Adult Female.—Distinguished from the male by being somewhat smaller, and by having the upper-parts faintly and irregularly barred with rufous-buff and black, and the chest-feathers margined externally with brown.

Range.—West Africa, from the Loango Coast northwards to Senegambia.

a. The three following species have the *breast and flanks whitish-buff, uniformly barred with black.*

V. THE GREY FRANCOLIN. FRANCOLINUS PONDICERIANUS.

Tetrao pondicerianus, Gmel. S. N. i. pt. ii. p. 760 (1788).
Francolinus pondicerianus, Steph. in Shaw's Gen. Zool. xi. p. 321 (1819); Ogilvie-Grant, Cat. B. Brit. Mus. xxii. p. 141 (1893).

Perdix orientalis, J. E. Gray, Ill. Ind. Zool. pl. 56, fig. 2 (1830-32).
Ortygornis pondicerianus, Hume and Marshall, Game Birds of India, ii. p. 51, pl. (1879); Oates, ed. Hume's Nests and Eggs Ind. B. iii. p. 435 (1890).

Adult Male and Female.—General colour above a mixture of chestnut and brown, barred with buff; below whitish-buff, closely barred with narrow wavy black bars. The *male* has a pair of sharp spurs. Total length, 12·5 inches; wing, 5·8; tail, 3·5; tarsus, 1·6.

Range.—South-western Asia, from Eastern Arabia and South Persia to India and Ceylon. Amirante and Mascarene Islands [introduced].

Habits.—From Mr. Hume's excellent account of the Grey Partridge, as it is called in India, the following notes on its habits are extracted:—

"Dry warm tracts, interspersed with scrub or low grass jungle, in the neighbourhood of cultivation, are what it specially affects, and the stunted acacia or wild date thickets or prickly pear hedges, that so often encircle our villages, are favourite haunts. So, too, are the hedges in some parts of the country enclosing every field, the bush-clad banks of nallas and broken ground, and ravines running down to rivers, more or less thinly or thickly studded with low catechu, acacia, or other scrub.

"Morning and evening they will be found in the fields or pecking about on the highways and byeways, but their homes are in the scrub, or in low thorny trees, in which many of them, in such localities, roost, and on which they may be found perching, at times, at almost any hour of the day.

"But provided the locality be dry and warm and the ground broken, no want of scrub or cultivation, no lack of trees and hedges, seems to banish them. I have shot them in the most desolate spots near the bases of the hills in Sind and on the

Mekran Coast, where there were no *traces* of vegetation at the time, and where, in the best of seasons only, a few straggling tufts of grass and desert plants are to be seen.

"The most noteworthy point about this species is its clear ringing, inspiriting call *kâ, kâ kateetur, kateetur*, which syllablize it as you will (and everyone has his own rendering), once heard, is never to be forgotten. Morning and evening the fields and groves re-echo with their cheery cry, and, during the spring and summer especially, it may be heard occasionally at all hours.

"They feed on grain of all kinds, grass seeds, and insects, especially white ants and their eggs, and on the young leaves of mustard, peas, and other herbs. Dig open an ant's nest in some scrub frequented by these birds, retire for ten minutes, and the chances are that on your return you find half a dozen Greys busy at the nest.

"'They run very swiftly and gracefully; they seem to glide rather than run, and the native lover can pay no higher compliment to his mistress than to liken her gait to that of the Partridge.

"It is often difficult to flush them, but when they rise it is with a true Partridge 'whir'; and their flight is swifter and stronger, and they will carry off more shot than our English bird.

"In many places they are to be found in pairs, but where they are really numerous, they often keep in regular coveys, a dozen rising within a small space if they are in ground in which they cannot run well."

Nest.—A slight hollow scratched out by the birds, generally in the shelter of scrub-jungle.

Eggs.—Six to nine in number; pointed ovals in shape; white, tinged with brownish-buff. Average measurements, 1·3 by 1·03 inch.

VI. THE COQUI FRANCOLIN. FRANCOLINUS COQUI.

Perdix coqui, Smith, Rep. Exp. Centr. Afr. p. 55 (1836).
Francolinus subtorquatus, Smith, Ill. Zool. S. Afr. Birds, pl. 15 (1838); Sharpe, ed. Layard's B. S. Afr. p. 600 (1884).
Francolinus coqui, Ogilvie-Grant, Cat. B. Brit. Mus. xxii. p. 143 (1893).
Francolinus stuhlmanni, Reichen. J. f. O. 1889, p. 270.

Adult Male.—Head chestnut-brown, shading into pale rufous or whitish on the throat; general colour of upper-parts a mixture of chestnut and buff, barred with blackish or dark grey; *back of the neck and under-parts white*, shading into buff on the belly, *all with wide regular black bars*. A pair of sharp spurs. Total length, 11 inches; wing, 5·5; tail, 3; tarsus, 1·5.

Adult Female.—Distinguished from the male by having black eyebrow stripes; the throat margined by a black band; and the *back of the neck and breast vinaceous-grey and dull chestnut, with white shaft-streaks*. No spurs.

Range.—This species has a wide range, being found over East, South, and South-west Africa.

Habits.—The habits of the Coqui Francolin are apparently much the same in all parts of its wide range.

Mr. T. Ayres writes: "These birds live in the open country, and are generally dispersed all over the Colony of Natal; they are to be found in coveys, like the Partridge in England; they roost on the ground in any convenient tuft of grass, and nestle all together. These birds would be numerous were it not for the burning of the grass, together with the hawks, wild cats, and snakes, which abound here and are their mortal enemies." This Francolin is extremely difficult to flush, and without the assistance of dogs is consequently seldom seen. Its call-note is shrill, but not unpleasant, and is mostly heard in the early morning and towards evening. Like the rest of its kind, its food consists of small bulbous roots, seeds, berries, and insects, and its flesh is excellent.

VII. HUBBARD'S FRANCOLIN. FRANCOLINUS HUBBARDI.

Francolinus hubbardi, Grant, Bull. B.O. Club, iv. p. xxvii. (1895).

Adult Male and Female.—Similar to *F. coqui*, but having the entire breast uniform buff, *without any black bars*. Total length, 10 inches ; wing, 5·6.

Range.—Nassa district, Victoria Nyanza.

VIII. SCHLEGEL'S FRANCOLIN. FRANCOLINUS SCHLEGELI.

Francolinus schlegelii, Heugl. J. f. O. 1863, p. 275 ; id. O.n. N. O.-Afr. ii. p. 898, pl. xxx. (1873) ; Ogilvie-Grant, Cat. B. Brit. Mus. xxii. p. 145 (1893).

? *Francolinus buckleyi,* Shelley MS. ; Ogilvie-Grant, Ibis, 1892, p. 41.

Adult Male.—Differs from the male of *F. coqui* chiefly in having the shoulders, wing-coverts, and outer webs of the secondary flight-feathers uniform light red.

Two female specimens (*F. buckleyi*) from Accra, now in the British Museum, which originally formed part of the Shelley collection, may prove to be the females of this species. They differ from the *female* of *F. coqui* in having the black stripes over the eye and round the throat nearly obsolete, the basal part of the inner primary and secondary flight-feathers chestnut, and the upper-parts greyer.

Range.—Bongo, Equatorial Africa, and perhaps extending to Accra on the West Coast.

b. In the following species the breast and flanks are *not* whitish buff, uniformly barred with black.

IX. THE RING-NECKED FRANCOLIN. FRANCOLINUS STREPTOPHORUS.

Francolinus streptophorus, Ogilvie-Grant, Ibis, 1891, p. 126 , id. Cat. B. Brit. Mus. xxii. p. 145, pl. 1 (1893).

Adult Male.—General colour above brown, below buff ; the sides of the head mostly bright chestnut ; eyebrow stripe, and another stripe across the hinder part of the cheek, and the throat white; from all the following species of this section it differs in having *a wide band of feathers barred alternately with black and white round the neck*. Total length, 11 inches; wing, 6·2; tail, 2·7; tarsus, 1·55. No spurs.

Adult Female.—Differs from the male in being rather smaller and in having the upper-parts barred with buff and the wing-coverts spotted with the same colour.

Range.—Central East Africa; southern foot of Mount Elgon and Masai-land. This fine species, recently obtained for the first time by Mr. F. J. Jackson, was met with in the scrubby plains in the localities mentioned above.

X. SMITH'S FRANCOLIN. FRANCOLINUS SEPHÆNA.

Perdix sephæna, Smith, Rep. Exped. Cent. Afr. p. 55 (1836).
Francolinus pileatus, Smith, Ill. Zool. S. Afr. Birds, pl. 14 (1838); Sharpe, ed. Layard's B. S. Afr. p. 593 (1884); Ogilvie-Grant, Cat. B. Brit. Mus. xxii. p. 146 (1893).

Adult Male.—Not unlike *F. streptophorus*, but the feathers surrounding the neck are *dark chestnut, edged on either side with white or buff*. The breast and under-parts are *without* chestnut spots. Total length, 12 inches; wing, 6·5; tail, 3·8; tarsus, 1·7. A pair of sharp spurs.

Adult Female.—Differs from the *male* in having the upper-parts covered with narrow wavy bars of buff and lines of black. No spurs.

Range.—South Africa, extending in the east from the Marico River and the Transvaal to the Zambesi, and westwards to northern Damara-land.

Habits.—This species inhabits the forest-clad hillsides and bush country, and is chiefly met with in the open glades. When flushed it generally perches on one of the higher branches, and with elevated crest inspects the movements of its pursuers. It is a somewhat rare bird, occurring in coveys, and appears to be very similar to *F. coqui* in its general habits, so far as one can gather from the scanty notes on the subject.

XI. GRANT'S FRANCOLIN. FRANCOLINUS GRANTI.

Francolinus granti, Hartl. P. Z. S. 1865, p. 665, pl. 39, fig. 1;
    Ogilvie-Grant, Cat. B. Brit. Mus. xxii. p. 148 (1893).
Francolinus schoanus, Heugl. Orn. N. O.-Afr. ii. p. 891, pl. xxix.
    fig. 2 (1873).
Francolinus ochrogaster, Hartl. J. f. O. 1882, p. 327.

This species is a smaller representative of *F. sephæna* in East Africa, and both sexes differ only from those of the latter bird in being less in size.

Male.—Total length, 11 inches; wing, 5·5; tail, 3·5; tarsus, 1·7.

Female.—Somewhat smaller, and devoid of spurs.

Range.—East Africa, extending from about 5° S. to 10° N. lat. and inland to about 31° E. long.

XII. KIRK'S FRANCOLIN. FRANCOLINUS KIRKI.

Francolinus kirki, Hartl. P. Z. S. 1867, p. 827; Finsch and
    Hartl. Vög. Ost-Afr. p. 588, pl. x. fig. 1 (1870); Ogilvie-
    Grant, Cat. B. Brit. Mus. xxii. p. 149 (1893).

Adult Male and Female.—Closely resemble in plumage the male and female respectively of both *F. sephæna* and *F. granti*, and in size agree with the latter species; but they are easily distinguished from both by having *an oblong chestnut spot at the end of the shaft of most of the feathers of the breast and belly.*

Range.—East Africa, from the Rovuma River to Dar-es-Salaam and Zanzibar Island.

XIII. THE SPOTTED FRANCOLIN. FRANCOLINUS SPILOGASTER.

Francolinus spilogaster, Salvadori, Ann. Mus. Civ. Genov. vi.
    p. 541 (1888); Ogilvie Grant, Cat. B. Brit. Mus. xxii.
    p. 149 (1893).

Adult Male.—Exactly like the male of *F. kirki*, but larger. Wing, 6·5 instead of 5·7 inches.

Range.—North-east Africa; Harar. So far as we are aware, only one male specimen (the type) of this species is known, and it bears the same relationship to *F. kirki* that *F. sephæna* bears to *F. granti*.

XIV. THE WHITE-THROATED FRANCOLIN. FRANCOLINUS ALBIGULARIS.

Francolinus albogularis, Gray, List Gall. B. iii. p. 35 (1844); Ogilvie-Grant, Cat. B. Brit. Mus. xxii. p. 149, pl. ii. (1893).

Adult Male.—General colour above chestnut, blotched and barred with black on the back; greyer on the rump; throat white; under-parts uniform buff. Total length, 9 inches; wing, 5·2; tail, 2·2; tarsus, 1·4. A pair of sharp spurs.

Range.—W. Africa; Gambia, Casamanze.

This little Francolin is at present only known from a few specimens, and nothing whatever is known regarding its habits or nidification.

XV. HARRIS' FRANCOLIN. FRANCOLINUS SPILOLÆMUS.

Francolinus psilolæmus (*sic*), Gray, List Gallinæ Brit. Mus. p. 50 (1867).

Francolinus spilolæmus, Finsch and Hartl. Vog. Ost-Afr. p. 586 (1870); Ogilvie-Grant, Cat. B. Brit. Mus. xxii. p. 150, pl. iii. (1893).

Adult Male and Female.—General colour above umber-brown, blotched with black and barred with buff; below buff, with a V-shaped black mark near the extremity of each feather, and a blotch of chestnut on the outer web; flight-feathers mostly bright chestnut; *chin and throat white, with a round black spot near the tip of each feather*. Male with a moderate pair of spurs.

Male: Total length, 12·3 inches; wing, 6·4; tail, 2·9; tarsus, 1·7.

Female: A little smaller.

Range.—North-east Africa; Shoa.

XVI. RÜPPELL'S FRANCOLIN. FRANCOLINUS GUTTURALIS.

Perdix gutturalis, Rüppell, Neue Wirb. p. 13 (1835).
Francolinus gutturalis, Rüppell, Syst. Uebers, p. 103, pl. 40;
  Ogilvie-Grant, Cat. B. Brit. Mus. p. 151 (1893).

Adult Male and Female.—Like *F. spilolæmus*, but only the feathers at the *edges of the throat* are spotted with black; feathers of the chest chestnut, mottled with grey and buff along the shaft; the breast and under-parts buff, *striped with black along the shafts;* and the sides and flanks are heavily blotched with chestnut, and barred with blackish-brown.

Male: Total length, 12·5 inches; wing, 6·5; tail, 2·8; tarsus, 1·4.

Female: Rather smaller.

Range.—North-east Africa; Abyssinia, Bogos, and the Mountains of Somali-land.

Habits.—According to Mr. W. T. Blanford, who had many opportunities of observing the species in Abyssinia, they were generally met with in small coveys during the months of December, January, and February, and subsequently seen in pairs, generally amongst bushes in valleys, and not keeping to the rocky hillsides where Sharpe's Francolin (*F. sharpii*) was to be found. They were not seen in the pass, but were common around Senafé, and moderately so throughout the highlands. In July and August the flesh was sometimes so rank as to be scarcely eatable, doubtless from their having fed largely on *Coleoptera*, which then abounded; but in the winter months they were excellent. The call, he says, is very similar to that of the common English Partridge, to which the plumage also presents some resemblance, so that sportsmen often take them to be the same bird.

XVII. THE ULU FRANCOLIN. FRANCOLINUS ULUENSIS.

Francolinus uluensis, Ogilvie-Grant, Ibis, 1892, p. 44; id. Cat.
B. Brit. Mus. xxii. p. 151 (1893).

Adult Male.—Intermediate between the last species, *F. gutturalis*, and the next one, *F. africanus*. It differs from the former and resembles the latter in having a *triangular patch of white feathers, tipped with black*, on each side of the neck, and the black marking on the breast and belly *arch-shaped*, giving these parts a *spotted* appearance. From *F. africanus* it differs in having the inner webs of the primary flight-feathers *mostly chestnut*. Total length, 12 inches; wing, 6·5; tail, 2·9; tarsus, 1·6. A pair of sharp spurs.

Range.—East Africa; Ulu country.

The only known examples of this species, both males, were recently obtained by Mr. F. J. Jackson at a place called Machako's in the above-mentioned country. They are particularly interesting, since they supply the intermediate link between the two very distinct forms, *F. gutturalis* from Abyssinia and *F. africanus* from South Africa, and inhabit an intermediate geographical area.

XVIII. THE PEARL-BREASTED FRANCOLIN. FRANCOLINUS AFRICANUS.

Francolinus africanus, Steph. in Shaw's Gen. Zool. xi. p. 323
(1819); Ogilvie-Grant, Cat. B. Brit. Mus. xxii. p. 152
(1893).

Francolinus afer (Latham, nec Müll.), Sharpe, ed. Layard's
B. S. Afr. p. 595 (1884).

Adult Male and Female.—General colour of upper-parts, throat and chest very similar to that of the two last species; breast and under-parts pale buff or whitish, with *irregular arch-shaped*, black bars, producing a pearled or ocellated appearance; *a large patch of black and white barred feathers* on each side of the neck; *inner webs of the primary flight-feathers brown*, more or less mottled with rufous. Total length, 13 inches;

wing, 6·4 ; tail, 3 ; tarsus, 1·5. Male with a pair of rather blunt spurs.

Range.—Eastern South Africa ; Transvaal to Cape Colony.

Habits.—This species is chiefly found in stony elevations and on the sides of mountains, and is decidedly a high-ground bird. Though the flight is strong, it is rarely sustained for any great distance, but the birds are said to run so swiftly that a winged one is almost certain to escape, unless followed by a good dog. The strong, curved bill of this Francolin enables it to dig up with great ease the small bulbous roots and insects which form its chief food.

Eggs.—Six to eight in number, varying in colour from light-green, almost white, to greenish-brown, and minutely dotted with brown. Measurements, 1·6 by 1·2 inch.

XIX. FINSCH'S FRANCOLIN. FRANCOLINUS FINSCHI.

Francolinus finschi, Bocage, Orn. Angola, p. 406 (1881).

Adult Male and Female.—General colour of the upper-parts as in the preceding species, but easily distinguished by its larger size and by having the entire sides of the head, as well as the sides and base of the throat, *uniform reddish-buff* : chin and middle of throat pure white ; rest of under-parts dusky-grey, shading into golden-buff on the belly, and spotted with chestnut. Bill very strong. *Male* with spurs. Total length, 14·5 inches; wing, 6·5 ; tail, 3·2 ; tarsus, 1·65.

Range.—South-western Africa ; Benguela.

Nothing is known of this rare Francolin, of which only a few examples, now in the Lisbon Museum, have been obtained at Caconda in the above-named country.

XX. THE CHESTNUT-NAPED FRANCOLIN. FRANCOLINUS CASTANEICOLLIS.

Francolinus castaneicollis, Salvadori, Ann. Mus. Civ. Genov. xxvi. p. 542 (1888) ; Ogilvie-Grant, Ibis, 1890, p. 350, pl. xi.; id. Cat. B. Brit. Mus. xxii. p. 153 (1893).

Only a single *adult female* of this rare species has so far been obtained and is preserved in the Turin Museum. It is a very distinct form, most nearly allied, perhaps, to *F. finschi*, which it resembles in having the *throat pure white*, neither spotted with black, nor margined by a black line. It is easily distinguished from *F. finschi* by having the forehead black, clothed with sharp rigid feathers; the nape chestnut; the feathers of the upper back chestnut, variegated with black and margined with grey. Total length, 12 inches; wing, 6·6; tarsus, 1·7.

Range.—North-east Africa: Lake Ciar-Ciar, Shoa.

Although this species is said to be common at the above locality, no additional specimens have so far been obtained and the male is still unknown.

XXI. LEVAILLANT'S FRANCOLIN.   FRANCOLINUS LEVAILLANTI.

Perdix levaillantii, Valenc. Dict. Sci. Nat. xxxviii. p. 441 (1825).
Perdix vaillanti, Temm. Pl. Col. v. pl. 33 [No. 447] (1829).
Francolinus levaillanti, Smith, Ill. Zool. S. Afr. pl. 85 (1843); Sharpe, ed. Layard's B. S. Afr. p. 596 (1884); Ogilvie-Grant, Cat. B. Brit. Mus. xxii. p. 154 (1893).

Adult Male and Female.—General colour above brown, blotched with black and barred with buff: breast bright chestnut; belly buff, blotched with chestnut. Easily distinguished from all the preceding species by having the rufous-buff throat *circumscribed by a well-marked black and white line*. From the following species it differs conspicuously in the disposition of the black and white stripes which ornament the sides of the head. The upper ones commence behind the nostril, pass above the eyes and ear-coverts, *surround the crown, and, uniting on the nape*, are continued down the middle of the back of the neck, and end in a patch of black and white feathers. The lower pair margin the chin and throat and form a crescent-shaped patch on the middle of the chest. Flight-feathers chestnut. Bill very strong. Total length, 13 inches; wing, 6·5; tail, 2·6; tarsus, 1·5.

Range.—South Africa; Transvaal, Orange Free State, Natal, Cape Colony.

Habits.—This remarkably handsome species, generally called the "Red-wing" in South Africa, is met with chiefly in the more secluded valleys forming the beds of the streams which flow between the high mountain ranges. Tufts of coarse grass and rushes are its favourite cover, from which it is extremely difficult to flush it, and birds of this species lie so close that Mr. Layard tells us he has on several occasions actually parted the grass under the pointer's nose to allow them to rise! If flushed a second time and well marked down, they may often be caught by the hand, as they will hardly rise again. In dry weather they keep so close to the dense palmiet that it is impossible to get them out. In the eastern districts they affect the damper parts of the hillsides and do not frequent the morasses.

Mr. T. Ayres says that their call is harsh and loud, and generally uttered morning and evening. The flight is rapid and strong, and they generally manage to settle out of sight behind some hillock or bush, where they are not easily found a second time.

Eggs.—Rather larger and redder in colour than those of *F. africanus*, described above.

XXII. THE GARIEP FRANCOLIN.   FRANCOLINUS GARIEPENSIS.

Francolinus gariepensis, Smith, Ill. Zool. S. Afr. pls. 83 [male] and 84 [female] (1849); Ogilvie-Grant, Cat. B. Brit. Mus. xxii. p. 155 (1893).

Adult Male and Female.—Like *F. levaillanti* in general colouring, but, among other differences, may be noted the position of the upper black and white stripes, which *do not* meet on the nape, but, passing along the sides of the neck, *unite with the lower line* which borders the throat; the chest and under-parts are rich buff, heavily blotched on one or both webs with dark chestnut. Bill strong. Total length, 13.5 inches; wing, 6.8; tail, 3.1; tarsus, 1.6. The female is generally devoid of spurs, but some examples have a blunt knob on one or both legs.

Range.—Eastern portion of South Africa, west of the Drakensberg Mountains.

Nest.—Placed in a depression in the ground among rough grass in some dry spot not far from water (*Ayres*).

Eggs.—Rather short and peg-top-shaped; tawny, spotted all over with dark brown. Measurements, 1·5 by 1·05 inch.

XXIII. BÜTTIKOFER'S FRANCOLIN. FRANCOLINUS JUGULARIS.

Francolinus gariepensis, Strickland and Sclater (*nec* Smith),
    Cont. Orn. 1852, p. 157.
Francolinus jugularis, Büttikofer, Notes Leyd. Mus. xi. pp. 76,
    77, pl. iv. (1889); Ogilvie-Grant, Cat. B. Brit. Mus. xxii.
    p. 156 (1893).

Adult Male and Female.—A paler western form of *F. gariepensis*, with the general colour of the plumage paler, especially on the lower breast and belly, which are pale buff, with only a few chestnut and blackish spots. The black and white bands encircling the throat generally form a *well-marked patch of black and white feathers on the front of the neck*.

Range.—Western South Africa, from Great Namaqua-land northwards to Angola.

Habits.—Andersson only met with this Francolin on the high table-lands, and always on grassy slopes sprinkled with dwarf bush. Though often very abundant and generally found in coveys of six to eight birds, it lies so very close, after having been once or twice flushed, that it is almost impossible to find again even with the assistance of dogs.

XXIV. SHELLEY'S FRANCOLIN. FRANCOLINUS SHELLEYI.

Francolinus gariepensis, Finsch and Hartl. Vög. Ost-Afr. p. 582
    (1870; *nec* Smith).
Francolinus shelleyi, Ogilvie-Grant, Ibis, 1890, p. 348; id. Cat.
    B. Brit. Mus. xxii. p. 157, pl. vi. (1893).

Adult Male and Female.—Differ from both *F. gariepensis* and *F. jugularis* in having the breast and belly *white, with V-shaped* *black bars*; it resembles the latter species in having a patch of black and white feathers at the base of the throat. Total length, 13 inches; wing, 5·8; tail, 3·4; tarsus, 1·5.

Range.— Eastern South Africa; Natal, Swaziland, to Mashona-land and Nyasa-land, and probably north to Zanzibar.

Habits.— Mr. Ayres writes: "This is the commonest of the Francolins on the Umvuli River, where it frequents the grassy and rocky slopes of the adjacent ranges. On the 7th of September a nest was found with three eggs; it was placed in a slight excavation in the ground, amongst high dry grass, and was lined with soft, half-decayed grass bents, mixed with a few feathers. The eggs were slightly incubated."

XXV. THE ELGON FRANCOLIN. FRANCOLINUS ELGONENSIS.

Francolinus elgonensis, Ogilvie-Grant, Ibis, 1891, p. 126; id. Cat. B. Brit. Mus. xxii. p. 157, pl. v. (1893).

Adult Female.— Allied to the last three species, but most nearly to the last-mentioned *F. shelleyi*. From all it is easily distinguished by having the *nape and upper back rufous-chestnut, with rounded spots of black*, and the sides of the face and neck between the black and white stripes clear buff; the ground-colour of the upper-parts mostly rich black; belly and under-parts black, tipped and barred with buff mixed with rufous. Total length, 12 inches, wing, 6·9; tail, 3·2; tarsus, 1·8.

Range.—Central East Africa; Mount Elgon.

Mr. F. J. Jackson, the discoverer of this fine species of Red-winged Francolin, obtained the only example known at an elevation of 11,000 feet. It was shot out of a flock of four, and he believes it to be the same species that he saw on the Mau escarpment at a height of 9,000 feet.

XXVI. THE INDIAN SWAMP FRANCOLIN. FRANCOLINUS GULARIS.

Perdix gularis, Temm. Pig. et Gall. iii. pp. 401. 731 (1815); J. E. Gray, Ill. Ind. Zool. i. pl. 56, fig. 1 (1830-32).

Francolinus gularis, G. R. Gray, List Gall. Brit. Mus. iii. p. 34 (1844); Ogilvie Grant, Cat. B. Brit. Mus. xxii. p. 158 (1893).

Ortygornis gularis, Hume and Marshall, Game Birds of India. ii. p. 59, pl. (1879); Oates, ed. Hume's Nests and Eggs Ind. B. iii. p. 437 (1890).

Adult Male and Female.—Upper-parts narrowly barred with brown, black, and buff alternately; throat and fore-neck deep rust-colour; feathers of remainder of under-parts whitish or pale buff, margined on either web with a brown and black band; primary flight-feathers chestnut. Male with a pair of sharp spurs.

Male: Total length, 13 inches; wing, 7·2; tail, 4·1; tarsus, 2·4.

Female: Rather smaller.

Range.—Northern India; Terai region, skirting the southern bases of the Himalayas, from Pilibhit in the west to Sadiya in Eastern Assam, Cachar, and Tipperah.

The Kyah, or Grass Chukor, as it is also commonly called, is very locally distributed, occurring here and there throughout the Terai region mentioned above. Tickell gives the following account of its habits:—" It frequents wild places—a sandy soil with thickets of the jungle-rose, babool, and other thorns, alternating with beds of reeds and elephant-grass, and always near water. It resorts also to such cultivation as lies within half a mile or so of the river, such as 'surson' (mustard), 'urhur' (dal), and 'chunna' (gram), but shuns paddy-fields, grass-meadows, or tree-jungle. Very early of a morning, or in the evening, it may be stalked on foot and potted; but the proper way of shooting this bird is to penetrate the thickets and 'nul bun,' or reed-jungle, on elephants, and with a large force of beaters, when the 'Khyr' affords as good a day's sport as may be had in a Pheasant covert in England. When first beaten up, it rises freely, but well within shot, with a loud flurry and often a shrill cackle, and its size makes it an easy shot when the young sportsman becomes used to its sudden flush, and his elephant ceases to start at the sound. If missed, it does not fly far, but it is almost impossible to force it to take wing again ; and a winged bird runs at such a rate, doubling and skulking in the covert, that without good dogs it is hopeless to search for it."

Like the Grey Partridge, the " Kyah " is a very pugnacious bird. A writer in the " Bengal Sporting Magazine " says that almost every one examined will be found scarred and marked with wounds from fighting.

Nest.—Well constructed, of grass, placed in a depression on the ground.

Eggs.—Said to be ten to fifteen in number; broad ovals, pointed at the smaller end; brownish-buff, finely speckled with purplish-brown at the larger end. Average measurements, 1·45 by 1·2 inch.

C. All the remaining species of the genus are characterised by having no rows of buff spots on the primary flight-feathers and the feathers of the upper-parts *devoid* of white or buff shaft-streaks. In *F. erckeli* only, the last species of this group, *a few of the outer scapulars have buff shafts.*

XXVII. THE CLOSE-BARRED FRANCOLIN. FRANCOLINUS ADSPERSUS.

Francolinus adspersus, Waterhouse, in Alexander's Exp. ii. p. 267, pl. [immature bird] (1838); Bocage, Orn. Angola, p. 410 (1881); Ogilvie-Grant, Cat. B. Brit. Mus. xxii. p. 159 pl. vii. (1893).

Scleroptera adspersa, Gurney, ed. Andersson's B. Damaral. p. 247 (1872).

Adult Male and Female.—Above umber-brown, finely mottled with dirty white and black; mantle and *under-parts narrowly barred with black and white.* Easily distinguished by this character from *all* the following species. Total length, 12·6 inches; wing, 6·6; tail, 3·5; tarsus, 1·7. The *male* is provided with a pair of long sharp spurs.

Range.—Western South Africa, extending from the Orange River to the Cunene River and inland to Lake Ngami.

Habits.—The Close-barred Francolin is, according to Andersson, one of the commonest species in Damara and Great Namaqua Land, where in favourable seasons the coveys often contain ten to fourteen birds. It generally frequents the banks of streams and perches much in trees, roosting among the branches at night, and retiring there during the heat of the day or on the approach of danger. This bird is always loath to fly unless very hard pressed, when it dives at once into the nearest thick tree and remains motionless; it generally prefers to escape by running with extraordinary swiftness. Its cry is extremely loud and harsh, and resembles "a succession of hysterical laughs, at first slow, but increasing in rapidity and strength, till they suddenly cease."

Eggs.—Cream-coloured. Measurements, 1·7 by 1·1 inch.

c. The following species have the *inner webs* of the primary flight-feathers either mostly pale buff, or brown, largely barred and mottled with chestnut or buff.

XXVIII. THE GREY-STRIPED FRANCOLIN. FRANCOLINUS GRISEOSTRIATUS.

Francolinus griseostriatus, Ogilvie-Grant, Ibis, 1890, p. 349, pl. x.; id. Cat. B. Brit. Mus. xxii. p. 160 (1893).

Adult Male.—Feathers of the upper-parts mostly dark chestnut, *margined on either side by a black and pearl-grey band; fore-neck* and *chest* very similarly marked; rest of under-parts buff, with wide dull rufous shaft-stripes; quills and tail chestnut, marked with black. A pair of sharp spurs.

Range.—West Africa; River Coanza.

A single male specimen of this fine Francolin is all that has so far been obtained.

XXIX. THE DOUBLE-SPURRED FRANCOLIN. FRANCOLINUS BICALCARATUS.

Tetrao bicalcaratus, Linn. S. N. i. p. 277 (1766).
Perdix senegalensis, Bonn. Tabl. Encyl. Méth. i. p. 212, pl. 93, fig. 2 (1791).
Perdix adansonii, Temm. Pig. et. Gall. iii. pp. 305, 717 (1815).
Francolinus bicalcaratus, Gray, List Gall. Brit. Mus. p. 33 (1844); Ogilvie-Grant, Cat. B. Brit. Mus. xxii. p. 160 (1893).

Adult Male and Female.—Plumage above brown, mottled with black, the feathers of the back and wings with a sub-marginal whitish band; throat white; *chest* and rest of under-parts pale buff, most of the feathers margined on each web with *chestnut* and with a *racket-shaped black mark* down the middle. Male with *two* pairs of spurs on each foot, the second (upper) pair being shorter and less pointed. Total length, 12·5 inches; wing, 6·8; tail, 2·5; tarsus, 2·1.

Range.—West Africa, from the River Niger to the Mogador Coast and Cape Blanco, Marocco.

Eggs.—Uniform greenish stone-colour. Average measurement, 1·75 by 1·4 inch.

XXX. CLAPPERTON'S FRANCOLIN. FRANCOLINUS CLAPPERTONI.

Francolinus clappertoni, Children, in Denham and Clapperton's Trav. App. xxi. p. 198 (1826); Ogilvie-Grant, Cat. B. Brit. Mus. xxii. p. 162 (1893).
Perdix clappertoni, Cretzschm. in Rüpp. Atl. p. 13, pl. 9 (1826).

Adult Male and Female.—Top of the head and ground-colour of the back and wings rufous or olive-brown, each feather of the mantle widely margined *all round* with whitish buff, and irregularly barred with buff; throat white; chest and rest of under-parts creamy-buff, with wide blackish-brown shaft-stripes down the middle of the feathers; some of the flank-feathers blotched with deep chestnut; greater part of the inner webs of the primary flight-feathers *uniform buff*. Male with *two* pairs of spurs. Total length, 14 inches; wing, 7·4; tail, 3·1; tarsus, 2·3.

Range.—The Soudan; Kordofan, Darfur, Bornu.

Habits.—Very few examples of this rare species have been obtained, and the only notes on its habits are those given by the discoverers, who say that it frequented sand-hills covered with low shrubs, and was very difficult to procure, owing to the speed with which it ran.

XXXI. GEDGE'S FRANCOLIN. FRANCOLINUS GEDGII.

Francolinus gedgii, Ogilvie-Grant, Ibis, 1891, p. 124; id. Cat.
    B. Brit. Mus. xxii. p. 163 (1893); Sharpe, Ibis, 1892, p.
    551, pl. xiv.

Adult Male.—Like *F. clappertoni*, but has the top of the head and ground-colour of the rest of the upper-parts *very dark brown*; the white margins to the feathers are *very narrow* and confined to the *sides of the webs*. A pair of blunt spurs. Total length, 12 inches; wing, 7·5; tail, 2·8; tarsus, 2·4.

Range.—Central East Africa; Elgon Plains.

The only specimen as yet obtained was a male shot by Mr. Ernest Gedge, who accompanied Mr. F. J. Jackson on his journey to Uganda, and after whom it has been named. It is reported to be fairly common on the Elgon Plains, but is apparently very local, as it was not met with anywhere else.

XXXII. HARTLAUB'S FRANCOLIN. FRANCOLINUS HARTLAUBI.

Francolinus hartlaubi, Bocage, J. Sci. Lisb. ii. p. 350 (1869);
    id. Orn. Angola, p. 408 (1881); Ogilvie-Grant, Cat. B.
    Brit. Mus. xxii. p. 163 (1893).

We have never seen an example of this species, which appears to be nearly allied to *F. clappertoni*, but, according to Professor

Barboza du Bocage, it differs in having the upper-parts brown, *spotted* with fulvous and black; the inner webs of the primary flight-feathers *pale brown, mottled with buff* on the edges; the tail-feathers *blackish, barred and edged with white;* the breast white, strongly striped with brownish-black. The types are an *immature* male and female, in the Lisbon Museum.

Range.—Western South Africa; Mossamedes.

XXXIII. HEUGLIN'S DOUBLE-SPURRED FRANCOLIN.
FRANCOLINUS ICTERORHYNCHUS.

Francolinus icterorhynchus, Heugl. J. f. O. 1863, p. 275; id. 1864, p. 27; id. Orn. N. O.-Afr. ii. p. 894, pl. 29, fig. 1 (1873); Ogilvie-Grant, Cat. B. Brit. Mus. xxii. p. 163 (1893).

Adult Male and Female.—Allied to *F. clappertoni*, but the upper-parts are brown, finely mottled with black and buff, apparently much the same as in *F. hartlaubi;* the inner webs of the primary flight-feathers brown, barred with rufous-buff; feathers of the under-parts buff, strongly striped down the middle with brownish-black, and *often* with a detached oval, or round black spot at the end; tail blackish-brown with irregular *rufous-buff bars*. *Male* with two pairs of spurs on the feet; *female* with one or two pairs of blunt knobs. Total length, 13·5 inches; wing, 6·8; tail, 3·2; tarsus, 2·2.

Range.—North Central Africa; Bongo, Djur, Kosanga, and the country to the west of the Albert Nyanza.

XXXIV. SHARPE'S FRANCOLIN. FRANCOLINUS SHARPII.

Francolinus clappertoni, Des Murs (*nec* Children), in Lefebvre's Voy. en Abyss. p. 146, pl. xii. (1845-50).

Francolinus rueppellii, Gray, et auct.; Blanf. Geol. and Zool. Abyss. p. 425 (1870).

Francolinus sharpii, Ogilvie-Grant, Ibis, 1892, p. 47; id. Cat. B. Brit. Mus. xxii. p. 164 (1893).

Adult Male and Female.—Allied to *F. clappertoni* and *F. gedgii*, but the inner webs of the primary flight-feathers are brown, with longitudinal or transverse bars of rufous and buff; and (this is the chief difference) the feathers of the breast are *dark brown, narrowly margined all round with white*. *Male* with two pairs of long sharp spurs. Total length, 13·5 inches; wing, 7·3; tail, 3·3; tarsus, 2·3.

This species has been constantly confounded with its more western ally, *F. clappertoni*, from which it is really very distinct.

Range.—North-east Africa; Bogosland, Abyssinia, and Shoa.

This Francolin is met with at lower elevations than some of the other Abyssinian species, and, according to Mr. Blanford, is commonest in some of the valleys at from 4,000 to 5,000 feet.

d. In the remaining species of the genus, the inner webs of the primary flight-feathers are *uniform dark brown*, sometimes slightly dotted with buff towards the marginal extremity.

XXXV. THE CAPE FRANCOLIN. FRANCOLINUS CAPENSIS.

Tetrao capensis, Gmel. Syst. Nat. i. pt. ii. p. 759 (1788).
Francolinus capensis, Steph. in Shaw's Gen. Zool. xi. p. 333 (1819); Ogilvie-Grant, Cat. B. Brit. Mus. xxii. p. 165 (1893).
Francolinus clamator, Temm.; Sharpe, ed. Layard's B. S. Afr. p. 591 (1884).

Adult Male and Female.—General colour above black, each feather with two or three concentric white lines running parallel with the margin; under-parts very similar to the upper, but with well-defined white shaft-stripes down the middle of the feathers; *throat spotted with black*.

Male with one or two pairs of rather blunt spurs. Total length, 16 inches; wing, 8·8; tail, 4·4; tarsus, 2·7.

Female.—Considerably smaller; wing, 8 inches.

Range.—South Africa; Cape Colony, extending to the Orange River: Robben Island [introduced].

This fine Francolin, commonly known in the colony as the Cape Pheasant, is easily recognised by its large size, and occurs throughout the maritime districts, "delighting," according to Mr. Layard, "in bushy kloofs and watercourses, from which it is driven with difficulty, owing to its habit of perching on branches, just out of the reach of dogs. It at all times prefers to escape by running, instead of flying; and on Robben Island, where it abounds, having been placed there some years ago, whole flocks may be chased for a mile or more in full view, without once taking wing. They usually, on these occasions, make for the rocks on the beach, and will run out to the farthest extremity, regardless of the surf breaking over them."

Eggs.—Eight to fourteen in number; olive-brown, occasionally spotted. Measurements, 1·9 by 1·5 inch.

XXXVI. THE NATAL FRANCOLIN. FRANCOLINUS NATALENSIS.

Francolinus natalensis, Smith, S. Afr. Journ. (2), p. 48 (1833): id. Ill. Zool. S. Afr. Birds, pl. 13 (1838); Sharpe, ed. Layard's B. S. Afr. p. 592 (1884); Ogilvie-Grant, Cat. B. Brit. Mus. xxii. p. 167 (1893).

Adult Male and Female.—Above brown, finely mottled with black and buff, and very similar to *F. icterorhynchus*; underparts *whitish*, with a more or less V-*shaped* or concentric *black bar on each feather; throat spotted with black.*

Male with one or two pairs of spurs; the upper one, when present, blunt. Total length, 13·5 inches; wing. 7·2; tail, 3·8; tarsus, 1·9.

Female.—Considerably smaller than the male; wing, 6·3. Sometimes with a pair of blunt spurs.

Range.—Eastern South Africa; Matabele-land, Transvaal. Swaziland, Natal.

Habits.—These birds inhabit wooded situations in the proximity of water, and are specially common in the dense underwood that abounds along the coast. Their habits are very similar to those of the last species, and their cry is said by Smith to resemble that of the Guinea Fowl.

XXXVII. HILDEBRANDT'S FRANCOLIN. FRANCOLINUS HILDEBRANDTI.

Francolinus hildebrandti, Cabanis, J. f. O. 1878, pp. 206, 243, pl. iv. fig. 2 [*female*]; Ogilvie-Grant, Ann. and Mag. N. H. (6), iv. p. 145 (1889); Hunter, in Willoughby's East Africa and its Big Game, App. i. p. 292 (1889); Ogilvie-Grant, Cat. B. Brit. Mus. xxii. p. 168 (1893).

Francolinus altumi, Fischer and Reichenow, J. f. O. 1884, p. 179, pl. ii. [*male*].

Adult Male.—Upper-parts very similar to those of *F. natalensis*, but *without* black shaft-streaks; under-parts white, with a *large heart-shaped* black spot near the extremity of each feather; throat *white, spotted with black*. Two pairs of spurs, the upper one blunt. Total length, 13·5 inches; wing, 7·5; tail, 3·8; tarsus, 2·2.

Adult Female.—Upper-parts as in the male; but the under-parts are dull brick-colour; throat pale rufous-buff; outer webs of flight-feathers *brown*, mottled with rufous-buff. One or two pairs of sharp spurs.

Range.—East Africa; Pangani River to the Kikuyu Country and Lake Naivasha.

This is one of the most interesting species of the group, and it can be easily understood that the two sexes being so different in plumage, they were at first described under different names. It has now, however, been conclusively proved that they are the male and female of one and the same bird (see Hunter, *l. s. c.*), yet, notwithstanding this, we observe, with some astonishment, that in Dr. Reichenow's latest work on the Birds of East Africa, the sexes are still regarded as distinct species!

If any further evidence were required, we need only refer to the parallel case of different sexes, found in the next species,

F. johnstoni, which is very nearly allied to *F. hildebrandti*, the females of the two species being almost indistinguishable.

XXXVIII. JOHNSTON'S FRANCOLIN. FRANCOLINUS JOHNSTONI.

Francolinus johnstoni, Shelley, Ibis, 1894, p. 24; Ogilvie-Grant, Cat. B. Brit. Mus. p. 559 (1893).

Adult Male.—Very like the male of *F. hildebrandti*, but the general colour of the upper-parts is darker, and the feathers of the under-parts have uninterrupted black shaft-stripes, instead of heart-shaped marks.

Adult Female.—Scarcely to be distinguished from the female of *F. hildebrandti*, but rather darker in colour.

Range.—South-east Africa; Nyasa-land.

This species was recently obtained at Zomba and on the Milanji Hills by Mr. A. Whyte, one of Mr. H. H. Johnston's collectors.

XXXIX. FISCHER'S FRANCOLIN. FRANCOLINUS FISCHERI.

Francolinus fischeri, Reichenow, J. f. O. 1887, p. 51; Ogilvie-Grant, Cat. B. Brit. Mus. xxii. p. 169 (1893).

We have never had an opportunity of examining this species, of which only a single *female* example is known. It appears to be much like the *females* of *F. hildebrandti* and *F. johnstoni*, but is said to differ in having the upper-parts pale brown, with rust-coloured shaft-stripes; the under-parts *clear ochre*, some of the breast-feathers being spotted with black at the tips, and *a clear spot at the extremity of the flight-feathers*, which have the outer webs *rust-red*, barred with dark brown. Total length, 12·8 inches; wing, 6·4; tail, 4; tarsus, 2.

Range.—Eastern Central Africa; Ussere, Wembaere Steppes.

XL. THE SCALED FRANCOLIN. FRANCOLINUS SQUAMATUS.

Francolinus squamatus, Cassin, P. Ac. Philad. viii. p. 321 (1857); Ogilvie-Grant, Cat. B. Brit. Mus. xxii. p. 169 (1893).

Francolinus petiti, Bocage, J. Sc. Lisb. vii. p. 68 (1879).
Francolinus modestus, Cabanis, J. f. O. 1889, p. 89.

Adult Male and Female.—Feathers of back of neck and mantle *reddish-brown, blotched with black* and *edged with white ;* rest of the upper-parts brown, with darker centres, and finely mottled with black ; under-parts brownish-buff, with the shaft of the feathers *only* dark.

Male with one or two pairs of spurs, the lower one long and sharp. Total length, 13·5 inches; wing, 7·3 ; tail, 3·5 ; tarsus, 2·1.

Female.—Somewhat smaller ; wing, 6·6 inches. No spurs.

Range.—West Africa ; Loango Coast and Gaboon.

XLI. SCHUETT'S FRANCOLIN. FRANCOLINUS SCHUETTI.

Francolinus schuetti, Cabanis, J. f. O. 1880, p. 351, 1881, pl. ii. ; Ogilvie-Grant, Cat. B. Brit. Mus. xxii. p. 170 (1893).

Adult Male.—Very similar to *F. squamatus*, but the feathers of the neck and mantle are more widely margined, and with *grey ;* the feathers of the middle of the breast and belly are *dull grey* round the margins, *shading into brown* towards the shaft. Total length, 13 inches ; wing, 6·9 ; tail, 3·4 ; tarsus, 2·1.

Adult Female.—Differs slightly from the *male* in having the feathers of the middle of the breast and belly margined with *pale buff.* Somewhat smaller ; wing, 6·6 inches.

Range.—West Africa, from Lunda, Angola ; East Africa, Lake Naivasha, Kilimanjaro district, 5,000 to 6,000 feet, Chaga district, and Pangani River.

XLII. THE AHANTA FRANCOLIN. FRANCOLINUS AHANTENSIS.

Francolinus ahantensis, Temm. Bijdr. tot de Dierk. i. p. 49, pl. 14 (1854); Büttikofer, Notes Leyd. Mus. vii. p. 231 (1885) ; xi. p. 126 (1889) ; Ogilvie-Grant, Cat. B. Brit. Mus. xxii. p. 171 (1893).

Francolinus ashantensis, Gray, List Gall. Brit. Mus. p. 51 (1867).

Adult Male and Female.—Upper-parts as in *F. squamatus*, but the feathers of the back of the neck are blackish, and margined with white on the sides only; under-parts brown, with a wide sub-marginal white band, edged on either side with blackish-brown.

Male: Total length, 14 inches; wing, 7·3; tail, 3·4; tarsus, 2.

Female: Smaller; wing, 6·7 inches.

Range.—West Africa; Gold Coast and Liberia.

Eggs.—Supposed to belong to this species are "reddish-brown, sprinkled with violet, and much paler towards the poles" (*Büttikofer*).

XLIII. JACKSON'S FRANCOLIN. FRANCOLINUS JACKSONI.

Francolinus jacksoni, Ogilvie-Grant, Ibis, 1891, p. 123, 1892, p. 51, pl. 1; id. Cat. B. Brit. Mus. xxii. p. 171 (1893).

Adult Male.—Top of the head, nape, and upper back dark brownish-chestnut, with white and mottled grey, and with black margins to the feathers; rest of upper-parts brown, washed with rufous on the outer wing-coverts and tail; under-parts *bright chestnut*, the feathers widely *margined on both webs with white*. A pair of strong spurs, and a supplementary blunt knob on the left foot. *Bill* and *feet coral-red*. Total length, 15·5 inches; wing, 9·1; tail, 5·2; tarsus, 2·8.

Range.—East Africa; Mianzini, Masai-land.

This is another of the new Francolins, by far the largest and finest, discovered by Mr. F. J. Jackson during his journey to Uganda. It is the largest bird of the genus, as well as one of the handsomest, and the discoverer of this fine novelty may well feel proud of it. Though common where it occurred, unfortunately only two specimens were preserved, both males; and the female is still unknown.

XLIV. ERCKEL'S FRANCOLIN. FRANCOLINUS ERCKELI.

Perdix erckelii, Rüppell, Neue Wirbelth. p. 12 (1835).
Francolinus erckelii, Des Murs, in Lefebvre's Voy. en Abyss. p. 144, pl. 11 (1845); Ogilvie-Grant, Cat. B. Brit. Mus. xxii. p. 172 (1893).
Francolinus erkelii, Auct. *passim*; Blanford, Geol. and Zool. Abyss. p. 423 (1870).

Adult Male.—General colour above greyish-brown, each feather of the back and wing-coverts margined with dark chestnut; *scapulars with whitish shaft-stripes*; under-parts *grey*, shading into whitish, with an *oblong dark chestnut, sometimes blackish, spot at the extremity of the shaft*; bill black; feet yellowish. Two pairs of equally developed, stout, sharp, spurs. Total length, 17 inches; wing, 8·5; tail, 4·8; tarsus, 2·4.

Adult Female.—Differs from the male in having the scapulars, tail-coverts, and tail-feathers more barred with wavy black and buff bars, and the feet without spurs.

Range.—North-east Africa; Bogosland and Eastern Abyssinia to Shoa, and westwards to Wogara, 2,500 to 11,000 feet.

THE BARE-THROATED FRANCOLINS. GENUS PTERNISTES.

Pternistes, Wagler, Isis, 1832, p. 1229.

Type, *P. nudicollis* (Bodd.).

The characters which distinguish this genus are the same as those of *Francolinus*, but *the throat is naked*, and the naked patch round the eye is large and conspicuous.

The nine species comprising this genus may be divided into two sections:—

A. Feathers of the back and scapulars with *dark-brown* or *black* shaft-stripes (species 1 to 7, pp. 136-140).

B. Feathers of the back and scapulars with *white* shaft-stripes (species 8 and 9, pp. 140-141).

A. *Feathers of the back and scapulars with dark-brown or black shaft-stripes.*

I. THE CAPE BARE-THROATED FRANCOLIN. PTERNISTES NUDICOLLIS.

Tetrao nudicollis, Bodd. Tabl. Pl. Enl. p. 11, No. 180 (1783).
Pternistes nudicollis, Sharpe, ed. Layard's B. S. Afr. p. 589 (1884); Ogilvie-Grant, Cat. B. Brit. Mus. xxii. p. 174 (1893).
Francolinus nudicollis, Auctorum, *passim.*

Adult Male.—General colour above brown; feathers of *neck black, margined on the sides with greyish-white;* mantle grey, with very wide black shaft-stripes; breast and under-parts *black, with white shaft-stripes;* feathers from the gape to the cheek black (freckled with *white* in *females* and *young*); naked skin round eye and on throat crimson; feet similarly coloured and with a pair of sharp spurs. Total length, 15·5 inches; wing, 7·9; tail, 3·7; tarsus, 2·4.

Adult Female.—Differs chiefly from the male in having the sides of the feathers of the lower breast and belly *rufous-brown* instead of black. Smaller; wing, 7·3 inches. No spurs.

Range.—South Africa; Transvaal and Cape Colony.

This species is met with in the wooded districts, and is common in many of the maritime parts of its range.

Eggs.—Rather round in shape; pinkish cream-colour, finely speckled all over with chalky-white. Measurements, 1·75 by 1·5 inch.

II. HUMBOLDT'S BARE-THROATED FRANCOLIN. PTERNISTES HUMBOLDTI.

Francolinus humboldti, Peters, MB. Akad. Berl. 1854, p. 134.
Pternistes humboldti, Sharpe, ed. Layard's B. S. Afr. p. 589 (1884); Ogilvie-Grant, Ann. Mag. N. H. (6), vii. p. 145 (1891); id. Cat. B. Brit. Mus. xxii. p. 176 (1893).

Francolinus (*Pternistes*) *leucoparæus*, Fischer and Reichenow, J. f. O. 1884, p. 263.

Adult Male.—Like *P. nudicollis*, but the feathers from the gape to the cheek are *white*; the black shaft-stripes on the feathers of the mantle *narrow*, scarcely extending beyond the shaft; and the breast and under-parts black (with a *white stripe* on either side of the shaft in *younger males*). Sides of the neck *black*. Bill, feet, and naked skin round the eye and on the throat blood-red. Total length, 14 inches; wing, 7·4; tail, 3·5; tarsus, 2·4.

Adult Female.—Differs chiefly in having the feathers from the gape to the cheek black and white; the sides of the neck *white, with a wide black band down the middle*; and the under-parts with traces here and there of white shaft-stripes. Smaller; wing, 7 inches. No spurs.

Range.—East Africa, from the Tana River to the Zambesi.

III. SCLATER'S BARE-THROATED FRANCOLIN. PTERNISTES AFER.

Tetrao afer, Müll. S. N. Suppl. p. 129 (1766).
Tetrao rubricollis, Gmel. S. N. i. pt. ii. p. 758 (1788).
Pternistes sclateri, Bocage, J. Sc. Lisb. i. p. 327, pl. vi. (1868).
Pternistes rubricollis, Auctorum, *passim*.
Pternistes afer, Ogilvie-Grant, Cat. Brit. Mus. xxii. p. 177 (1893).

Adult Male.—Like *P. nudicollis*, but the eyebrow-stripes and feathers between the gape and cheek are *pure white*; breast and under-parts *white, with a wide black band down the middle of each feather*. Feathers of upper-chest grey, finely dotted towards the extremity with black. Total length, 14 inches; wing, 7·1; tail, 3·1; tarsus, 2·2.

Adult Female.—Differs chiefly in having the feathers of the chest brownish-grey, with black shaft-stripes. Smaller; wing, 6·5 inches. No spurs.

Range.—South-western Africa; Mossamedes, Benguela, Angola.

IV. CRANCH'S BARE-THROATED FRANCOLIN. PTERNISTES
CRANCHI.

Perdix cranchii, Leach, in Tuckey's Narrat. Explor. River Zaire, App. p. 408 (1818).
Pternistes cranchii, Wagler; Ogilvie-Grant, Cat. B. Brit. Mus. xxii. p. 178 (1893).
Perdix punctulata, J. E. Gray, Ill. Ind. Zool. ii. pl. 43. fig. 3 (1833-4).
Pternistes lucani, Bocage, J. Sc. Lisb. vii. p. 68 (1879).

Adult Male.—General colour above umber-brown, finely mottled with black; *neck, mantle,* chest, and under-parts finely mottled with black and white, *the feathers of the breast and belly being widely margined on both webs with chestnut.* A pair of sharp spurs. Naked skin of throat, bill, and feet crimson-red. Total length, 14 inches; wing, 7·3; tail, 3·1; tarsus, 2·2.

Adult Female.—Differs somewhat from the male, having the feathers of the back of the neck *brown, edged with white, and the mantle brown like the back.* No spurs. Smaller; wing, 6·8 inches.

Range.—West Africa; Congo and Loango Coast, and extending to the Marungu Country, south-west of Lake Tanganyika.

V. BOEHM'S BARE-THROATED FRANCOLIN. PTERNISTES
BOEHMI.

Francolinus cranchii, Finsch & Hartl. (*nec* Leach), Vög. Ost-Afr. p. 579, pl. ix. (1870).
Pternistes böhmi, Reich. J. f. O. 1885, p. 465; Ogilvie-Grant, Cat. B. Brit. Mus. xxii. p. 179 (1893).

Adult Male and Female.—Like *P. cranchi*, but the feathers of the breast and belly have *black shaft-stripes.*

Range.—Central East Africa, east of Lake Tanganyika; Usui, Victoria Nyanza, Unyamuesi, Unyanyembe, and Ugogo.

This species, which represents *P. cranchi* in East Africa, was at first confounded with that species, but is really perfectly distinct.

VI. SWAINSON'S BARE-THROATED FRANCOLIN. PTERNISTES SWAINSONI.

Perdix swainsoni, Smith, Rep. Exp. Centr. Afr. p. 54 (1836).
Francolinus swainsoni, Smith, Ill. Zool. S. Afr. Birds, pl. 12 (1838).
Pternistes swainsoni, Sharpe, ed. Layard's B. S. Afr. p. 587 (1884); Ogilvie-Grant, Cat. B. Brit. Mus. xxii. p. 170 (1893).

Adult Male.—General colour of plumage umber-brown, finely dotted with black; most of the feathers of the breast and belly margined on the sides with chestnut; naked skin round eye and on throat rose-red; upper mandible black, lower red; feet and toes blackish, with a ruddy tinge. A pair of long sharp spurs. Total length, 14 inches; wing, 8; tail, 3·1; tarsus, 2·4.

Adult Female.—Differs in having most of the feathers of the breast and belly *devoid of chestnut margins*. No spurs. Smaller; wing, 6·9 inches.

Range.—South Africa; Matabele, Transvaal, and Damaraland.

Habits.—The habits of this species appear to be very similar to those of *Francolinus capensis*. In Matabele-land Mr. Buckley tells us that this bird is called "Pheasant" by the colonists, and is generally found in coveys which, like those of the other species of this genus, are extremely difficult to flush, and always prefer, if possible, to escape by running. They are generally met with in the neighbourhood of small streams, coming out into the open in daytime, and passing the night in the brushwood, where they roost in the trees. Their food, like that of the rest of their kind, consists chiefly of bulbs, seeds, berries, and insects.

Eggs.—Six and probably more in number; rather round; pinkish cream-colour, finely speckled all over with chalky-white. Measurements, 1·7 by 1·5 inch.

VII. REICHENOW'S BARE-THROATED FRANCOLIN. PTERNISTES RUFOPICTUS.

Pternistes rufopictus, Reichenow, J. f. O. 1887, p. 52; Ogilvie-Grant, Cat. B. Brit. Mus. xxii. p. 180 (1893).

Adult Male.—Feathers of the neck *white, with black and rufous-brown margins*. General colour above grey, with dark cross-bars; below white, with black shaft-streaks and broad rufous-brown margins; naked skin on the throat orange-yellow; bill reddish-brown; feet and toes dark brown. Total length, 15·6 inches; wing, 8·4; tail, 3·2; tarsus, 2·8.

Range.—Equatorial Africa; Wembaere Steppes, Ussambiro, and Unyoro, Nassa district, Victoria Nyanza.

Very few examples of this remarkably fine Bare-throated Francolin have as yet been obtained, and I have only recently been able to examine perfect specimens, though I had seen the heads and necks of two which were brought from Unyoro. The British Museum, however, has recently acquired three specimens from Nassa, presented by the Rev. G. Hubbard.

B. Feathers of the back and scapulars with white shaft-stripes.

VIII. GRAY'S BARE-THROATED FRANCOLIN. PTERNISTES LEUCOSCEPUS.

Perdix rubricollis, Cretzschm. (*nec* Gmel.), Rüpp. Atl. p. 44, pl. 30 (1826).

Francolinus leucoscepus, Gray, List Gallinæ Brit. Mus. p. 48 (1867).

Pternistes rubricollis, Blanf. Geol. and Zool. Abyss. p. 426 (1870).

Pternistes leucoscepus, Ogilvie-Grant, Cat. B. Brit. Mus. xxii. p. 181, pl. viii. fig. 1 (1893).

Adult Male and Female.—General colour above brown; feathers of the chest and breast *white, with dull brown margins;* those of the belly, sides, and flanks similar, but with dull chestnut margins; bill dusky; naked skin round eye and on throat orange-red, shading into yellow; feet dusky-red.

Male.—With a pair of sharp spurs, sometimes supplemented by a second blunt pair. Total length, 15 inches; wing, 8·1; tail, 3·6; tarsus, 2·5.

Female.—Somewhat smaller and devoid of spurs.

Range.—North-east Africa; Bogosland, Abyssinia, Shoa, and North Somali-land.

IX. CABANIS' BARE-THROATED FRANCOLIN. PTERNISTES INFUSCATUS.

Pternistes infuscatus, Cabanis, J. f. O. 1868, p. 413, and in V. d. Decken's Reis. iii. p. 44, pl. 14 (1869); Ogilvie-Grant, Cat. B. Brit. Mus. xxii. p. 182, pl. viii. fig. 2 (1893).

Adult Male and Female.—Like *P. leucoscepus*, but distinguished by having the feathers of the chest *brown*, shading into chestnut towards the base, and each with *narrow white* shaft-stripes forming a triangular white patch at the extremity, so that the predominating colour is *brown* instead of white.

Range.—This is a more southern representative of *P. leucoscepus*, and it is probable that the two forms intergrade in Somali-land. The typical form is found in East Africa from Mamboio northwards to Kilimanjaro, the Teita district, and Southern Somali-land.

THE LONG-BILLED FRANCOLINS. GENUS RHIZOTHERA.

Rhizothera, G. R. Gray, List Gen. B. 2nd ed. p. 79 (1841).

Type, *R. longirostris* (Temm.).

Differs chiefly from *Francolinus* in having only twelve tail-feathers. The tail is rather more than half the length of the wing; the first primary flight-feather is about equal to the tenth, the sixth slightly the longest. Bill very stout, long, and curved. Feet in both sexes provided with a pair of stout spurs. Sexes quite different.

Only two species are known

I. THE LONG-BILLED FRANCOLIN. RHIZOTHERA LONGIROSTRIS.

Perdix longirostris, Temm. Pig. et Gall. iii. pp. 323, 721 (1815) [male].

Tetrao curvirostris, Raffl. Trans. Linn. Soc. xiii. p. 323 (1822) [female].

Francolinus longirostris, J. E. Gray, Ill. Ind. Zool. ii. pl. 45, fig. 2 (1833-4).

Rhizothera longirostris, G. R. Gray; Kelham, Ibis, 1882, p. 4; Ogilvie-Grant, Cat. B. Brit. Mus. xxii. p. 183 (1893).

Adult Male.—Top of the head rich brown; general colour above chestnut, blotched with black, shading into grey, mixed with buff on the lower back and upper tail-coverts; sides of head and throat reddish-chestnut; neck, chest, and upper mantle *grey*; rest of under-parts rufous-buff. Total length, 14·6 inches; wing, 7·7; tail, 3·5; tarsus, 2·2.

Adult Female.—Differs from the male in having the neck and chest *rufous-chestnut*, and the lower back and upper tail-coverts *mostly buff*. Slightly smaller than the male.

Range.—Southern part of the Malay Peninsula, and extending to Sumatra and Borneo.

II. HOSE'S LONG-BILLED FRANCOLIN. RHIZOTHERA DULITENSIS.

Rhizothera dulitensis, Ogilvie-Grant, Bull. Brit. Orn. Club, iv. p. xxvii. (1895).

Adult Male.—Easily distinguished from the *male* of *R. longirostris* by having the whole chest and *breast* grey, and the rest of the under-parts white.

Adult Female.—Differs from the female of *R. longirostris* in its generally richer colouring and in having the general colour of the outer wing-coverts dark brown, with comparatively few buff markings.

Range.—Mount Dulit, Sarawak, Borneo, at an elevation of 4,000 feet.

A pair of this fine Francolin were obtained by Mr. C. Hose.

THE TRUE PARTRIDGES. GENUS PERDIX.

Perdix, Briss. Orn. 1. p. 219 (1760).

Type, *P. perdix* (L.).

Differs from the preceding genera in having the tail composed of either sixteen or eighteen feathers, nearly equal in length, the outer pair being only slightly shorter than the middle pair, which are more than half the length of the wing. The first primary flight-feather is intermediate in length between the seventh and eighth, and the fourth is slightly the longest. The feet are without spurs in either sex, and the plumage of both is alike or slightly different.

The four species may conveniently be divided into two groups, each containing two species.

A. Tail with eighteen feathers; chest and breast not barred with black (species 1 and 2, pp. 143-149).

B. Tail with sixteen feathers; chest and breast barred with black (species 3 and 4, pp. 150-151).

A. *Tail with eighteen feathers; chest and breast not barred with black.*

1. THE COMMON PARTRIDGE. PERDIX PERDIX.

Tetrao perdix, Linn. S. N. i. p. 276 (1766).

Perdix cinerea, Latham, et auctorum plurimorum: Dresser, B. Europe, vii. p. 131, pls. 474 and 475 (1878).

Perdix robusta, Homeyer and Tancré, Mitth. Orn. Ver. Wien. vii. p. 92 (1883); ix. pl. figs. 3-5 (1885).

Perdix perdix, Ogilvie-Grant, "Field," 21st Nov. 1891, and 9th April, 1892; id. Ann. and Mag. N. H. (6), xii. p. 62 (1893); id. Cat. B. Brit. Mus. xxii. p. 187 (1893).

Adult Male.—General colour above brownish-buff (washed with grey in birds from Northern Europe), with narrow, close-set, wavy cross-bars and lines of black; *lesser and median wing-*

coverts and scapulars blotched on the inner web with chestnut, and with *only* buff shaft-stripes (fig. 1). Top of the head brown, rest of the head, throat, and neck chestnut; breast grey, finely mottled with black, below which is a large horse-shoe-shaped chestnut patch; rest of under-parts whitish; first flight-feather with extremity rounded; feet horn-grey. Total length, 12·6 inches; wing, 6·2; tail, 3·5; tarsus, 1·7.

Adult Female.—Easily distinguished from the *male* by having the ground-colour of the *lesser and median wing-coverts and scapulars mostly black, with wide-set buff cross-bars*, in addition to the longitudinal buff shaft-stripe down the middle of each feather (figs. 2 and 3); and the chestnut patch on the breast small, or sometimes absent.

Immature examples of both sexes exhibit the characteristics of the adults, but may be recognised by having the first primary flight-feather *pointed* at the extremity instead of being rounded, and the feet *yellowish horn-colour*.

The *immature female* has generally a well-developed chestnut horse-shoe mark on the breast.

Range.—Europe and Western and Central Asia, extending in the west to Scandinavia and the British Isles, in the east to the Barabinska Steppes and Altai Mountains, and in the south to Northern Spain and Portugal, Naples, the Caucasus, Asia Minor, and North Persia.

As considerable interest attaches to the sexual differences in plumage in the Common Partridge, it may be worth while to republish here the substance of my articles on this subject which appeared in the "Field" quoted above.

In every text-book on ornithology which gives a description of the plumage in the male and female of the Common Partridge we find that the chief difference mentioned as distinguishing the two sexes is, that the *male* has a large chestnut horse-shoe-shaped mark on the lower breast, while in the *female* this marking is reduced to a few chestnut spots, or sometimes entirely absent. This character, as we first pointed out

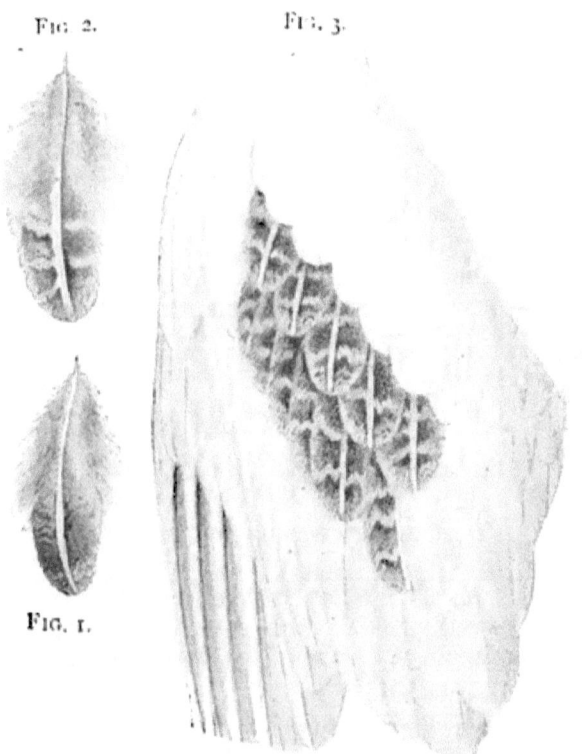

FIG. 1.—Median wing-covert of male Partridge. FIGS. 2 & 3.—Median wing-coverts of female Partridge.

in the "Field," is not to be depended on, for the great majority of young females—by which we mean birds of the year—have a well-developed chestnut horse-shoe, and in some, for instance birds from Leicestershire, it is quite as large and perfectly developed as in the majority of adult male birds. Young females from Norfolk and Suffolk are, however, *generally* exceptions to this rule, and, like the majority of old females, have merely a few chestnut spots on the middle of the lower breast, and in this part of England it is rare to meet with anything like a perfect horse-shoe in young birds of this sex, while examples may be found without a trace of chestnut, and are commonly known as birds with a white horse-shoe. As remarked above, the birds of the year, whether male or female, are easily distinguished from old birds by having the first flight-feather pointed instead of rounded at the extremity. The colour of the feet and toes is also, of course, a good character for distinguishing young birds from old ones in the earlier part of the season, but at the commencement of hard weather the yellowish-brown feet, denoting youth, having generally changed to bluish-grey, are perfectly similar to those of the adult, while the pointed first flight-feather is retained till the following autumn moult. The only reliable character for distinguishing the sexes at all ages, except in *very* young birds in their first plumage, is in the markings of the lesser and median wing-coverts and scapulars, the buff *cross-bars* in the female being an unmistakable mark, and quite sufficient to distinguish her from the male at a glance. It is now some years since we first drew attention to these rather important differences which had hitherto been entirely overlooked, and we may now safely say, that though many people, especially sportsmen, were at first disinclined to believe in this character being a sexual difference, and tested it severely, it has, so far, never been found to fail. To convince gamekeepers of these facts is in most cases a hopeless task—that the horse-shoe mark on the breast is a certain sign of the male is "bred in the bone," having been handed down as gospel for genera-

MOUNTAIN PARTRIDGE.

tions. One Scotch keeper in particular, at a place where we have enjoyed many a pleasant day's Partridge-shooting, rises before our mind, and the remembrance of this excellent and extremely obstinate soul always makes us smile. Often at lunch-time have we started him on the Partridge question, merely for the fun of hearing him argue, and stick to his opinion and that of his forefathers; and his politely incredulous smile on being shown by the help of a knife that some particular bird with a large horse-shoe mark really *was* a female by dissection, had to be seen to be appreciated. But there are some people who will never allow that they are mistaken, and as long as this good man remains, we may safely look forward to many a half-hour's amusement, though the dissection of numerous Partridges does not meet with our host's entire approval! It must be added that barren females are sometimes met with in more or less perfect male plumage. One barren female (by dissection), in the National Collection, has an enormous chestnut horse-shoe mark on the breast, while the wing-coverts have one web of each feather like that of the male, and the other barred as in the ordinary female. This, and one other example, are the only two that have come under our notice, though we have examined thousands of birds, and we may safely conclude that they are by no means common.

Varieties.—A curious rufous variety of the Common Partridge (Plate XII.) was first described under the name of *Perdix montana* by Brisson,* who believed it to represent a distinct species. This is not, however, the case, as every intermediate phase of plumage between the Common Partridge and the most extreme chestnut form can be found. The finest examples of this variety have the *whole* head and neck dull rust-red and the remainder of the plumage *dark chestnut*, except the thighs and lower part of the belly, which are whitish, as well as some bars and markings on the wing-coverts and scapulars. Brisson's

* Orn. i. p. 224, pl. xxi. fig. 2 (1760).

specimens were obtained in the mountains of Lorraine, but fine examples have also been procured in Northumberland, Cheshire, and Wiltshire in England, as well as from other localities, and there can be no doubt that this form is merely a sport of nature or accidental variety in which the chestnut colour pervades the whole plumage. Equally perfect examples of *both sexes* have been obtained. Grey, cream-coloured, and white examples of the Common Partridge are sometimes met with, but are by no means common, and generally prove to be birds of the year, probably because birds of peculiar plumage are generally shot down or killed by birds of prey, &c., while still young, being more conspicuous than their neighbours.

Nest.—A slight hollow in the ground, roughly lined with a few dry grasses, &c., and sheltered by rough grass, growing crops, or bushes.

Eggs.—Ten to fifteen, and sometimes as many as twenty; in shape pointed ovals; uniform pale olive-brown in colour. Average measurements, 1·4 by 1·1 inch.

SUB.-SP. *a*. THE MIGRATORY PARTRIDGE. PERDIX DAMASCENA.

Perdix damascena, Briss. Orn. i. p. 223 (1760), et auctorum.
Starna cinerea, var. *peregriana*, Tschusi u. de la Torre, Ornis. 1888, p. 250.

This sub-species or race is perfectly similar in plumage to the Common Partridge, and appears to be merely a smaller high-ground or Alpine form of the latter species, but the feet and toes, and apparently also the bill, are yellowish in the adult, instead of horn-grey. Total length, about 10 inches; wing, 5·9; tail, 3·1; tarsus, 1·4.

Large flocks of the Migratory Partridge visit the plains of Southern and Central Europe during the cold season, and are reported not to associate with birds of the common species. It seems probable that this supposed sub-species may prove to be founded on certain individuals of the Common Partridge which inhabit the higher elevations throughout its range, and get driven down in winter to the plains and valleys in search of food. The fact is, however, that very little is known about this migratory form; and, although its existence has been well-known for more than a century, very few examples have been obtained, and in none of those which have come under our notice have the colour of the feet and toes been recorded while the birds were *still fresh*. In all the examples we have examined, the feet certainly have the appearance of having been yellow or yellowish horn-colour, but all these are birds of the year, as may at once be seen by the pointed first primary quill.

In many parts of Scotland we have met with Partridges breeding on the lower moorland, and it is by no means uncommon to fall in with an isolated covey or two of these birds on the edges of a Grouse moor. Such "hill-birds" are, as a rule, smaller and more brightly coloured than the low-ground birds, and, when handled, seem to be about half the size of specimens from the southern counties of England. When measured, the differences in the length of the wing are comparatively trifling, but the average weight of the hill-birds is considerably less.

II. THE BEARDED PARTRIDGE. PERDIX DAURICA.

Tetrao perdix, var. *daurica*, Pall. Zoogr. Rosso-As. ii. p. 78 (1811).

Perdix (*Starna*) *cinerea*, var. *rupestris daurica*, Radde, Reise Ost-Sib. ii. p. 304, pl. xii. (1863).

Perdix barbata, Verr. and Desm. P. Z. S. 1863, p. 62, pl. ix.; Gould, B. Asia, vi. pl. 73 (1871); Prjev. in Rowley's Orn. Misc. ii. p. 422 (1877).

Perdix daurica, Ogilvie-Grant, Cat. B. Brit. Mus. xxii. p. 193 (1893).

Adult Male.—General appearance of *P. perdix*, but paler and greyer, and easily distinguished by having the feathers on the sides of the chin and throat considerably elongated, with dark shafts, *forming a beard*; the middle of the breast bright buff;

and the horse-shoe patch on the breast *black*. Total length, 11·5 inches; wing, 6; tail, 3·5; tarsus, 1·5.

Adult Female.—Differs from the male in having less buff on the breast, and the black patch on the breast much reduced in size, or absent. Smaller; wing, 5·6 inches.

Range.—North-eastern and Central Asia, extending north to Dauria, east to Amoorland, Manchuria, and the mountains near Pekin, west to Dzungaria and the Tian-shan Mountains, and south to the sources of the Yangtze-kiang.

B. *Tail with sixteen feathers; chest and breast barred with black.*

III. MRS. HODGSON'S PARTRIDGE.   PERDIX HODGSONIÆ.

Sacfa hodgsoniæ, Hodgson, J. As. Soc. Beng. xxv. p. 165, pl. (1857).
Perdix hodgsoniæ, Gould, B. Asia, vi. pl. 74 (1857); Hume and Marshall, Game Birds of India, p. 65, pl. (1879); Oates, ed. Hume's Nests and Eggs Ind. B. iii. p. 438 (1890); Ogilvie-Grant, Cat. B. Brit. Mus. xxii. p. 193 (1893).

Adult Male and Female.—Forehead with a black and a white band, the fore-part of the crown chestnut; a rufous-chestnut collar; mantle grey, shading into brownish-grey on the rest of the upper-parts, and all barred with black and rufous-buff; wings very similar to those of the male of *P. perdix*, but brighter; cheeks, chin, and middle of throat white, the feathers of the latter rather long, but shorter than in *P. daurica;* a *large* black patch covering the hinder part of the cheeks and side of the throat; under-parts white, barred with black, and with *a large black patch on the middle of the breast.* Total length, 11·5 inches; wing, 6·2; tail, 3·6; tarsus, 1·7.

Range.—Southern Tibet, just extending into Northern India, in Cashmere, Gurhwal, Kumaon, and Sikhim.

Habits.—This species inhabits the desolate hillsides and passes, covered with stones and rocks, at elevations varying from about 12,000 to 19,000 feet, where the only vegetation is patches of mossy herbage. Very little is known either of its habits or mode of nesting, but they appear to be very similar to those of the Common Partridge, and the flight is said to be identical.

Eggs.—Much like those of *P. perdix*; pale clay-brown, slightly tinged with a reddish-brown towards the poles. Measurements, 1·77 by 1·2 inch.

IV. PRJEVALSKY'S PARTRIDGE. PERDIX SIFANICA.

Perdix sifanica, Prjevalsky, Mongolia, ii. p. 124 (1876); id. in Rowley's Orn. Misc. ii. p. 423 (1877); Ogilvie-Grant, Cat. B. Brit. Mus. xxii. p. 195 (1893).

Adult Male and Female.—Like *P. hodgsoniæ*, but smaller, and differ chiefly in having the black patch on the hinder part of the cheek and side of the throat *much smaller*, and mixed with chestnut above; *no* black patch on the middle of the breast, all the feathers being white, barred with black like the rest of the under-parts.

Male: Total length, 10·5 inches; wing, 5·9; tail, 3·2; tarsus, 1·6.

Female: Somewhat smaller; wing, 5·5 inches.

Habits.—This species, Prjevalsky tells us, was met with principally in the rhododendron thickets in the Alpine regions of Kansu, where the mountains were covered with small tufts of *Potentilla tenuifolia*. It was also met with in the plains, which are, however, at an elevation of 10,000 feet above the sea-level.

THE MADAGASCAR PARTRIDGES. GENUS MARGAROPERDIX.

Margaroperdix, Reichenb. Av. Syst. Nat. p. xxviii. (1852).

Type, *M. madagascariensis* (Scop.).

Tail about half the length of the wing and composed of

twelve feathers, wedge-shaped, the middle pair of feathers being considerably longer than the next pair; the first primary flight-feather is intermediate in length between the sixth and seventh, and not much shorter than the longest. The tarsus is not provided with a spur in either sex, and the plumage is entirely different in the two sexes.

Only one rather small species is known.

I. THE MADAGASCAR PARTRIDGE. MARGAROPERDIX MADAGASCARIENSIS.

Tetrao madagarensis, Scop. (*ex* Sonnerat), Del. Flor. et Faun. Insubr. pt. ii. p. 93 (1786).

Perdix striatus, Lath.; Temm. Pl. Col. v. pl. 39 [No. 82] (1823).

Margaroperdix striatus, Auctorum, *passim;* Grandidier and Milne-Edwards, Hist. Madagas. Ois. i. p. 487, pls. 199–201A (1885).

Margaroperdix madagascariensis, Ogilvie-Grant, Cat. B. Brit. Mus. xxii. p. 196 (1893).

Adult Male.—General colour of upper-parts reddish-brown, with whitish shaft-stripes, and mostly with rufous or buff cross-bars; a line of black feathers with whitish shaft-stripes down the middle of the head; sides of head and throat black, with white stripes over the eye and along the sides of the throat; fore-neck and middle of chest chestnut; sides grey; middle of breast and belly black, with oval white spots; sides and flanks mostly chestnut; tail black, barred with reddish-white; bill black. Total length, 10 inches; wing, 5·1; tail, 2·7.

Adult Female.—General colour above black, mixed with olive-brown, with pale shaft-stripes and bars as in the male; throat, sides of the head, and under-parts mostly rufous-buff, the latter with concentric black lines on each feather; sides and flanks barred with black.

Range.—Island of Madagascar.

THE INDIAN BUSH-QUAILS. GENUS PERDICULA.

Perdicula, Hodgs. Beng. Sport. Mag. ix. p. 344 (1837).

Type, *P. asiatica* (Lath.).

Tail composed of twelve feathers, rather feeble, but much stiffer than the upper tail-coverts; less than half the length of the wing. First flight-feather intermediate in length between the seventh and ninth; fourth slightly the longest. Tarsus armed in the male with a blunt wart-like spur. Plumage of sexes different.

Only two very small species are known.

1. THE JUNGLE BUSH-QUAIL. PERDICULA ASIATICA.

Perdix asiatica, Lath. Ind. Orn. ii. p. 649 (1790).
Perdix cambayensis, Temm. Pl. Col. v. pl. 41 [No. 447] (1828).
Coturnix pentah, Sykes; J. E. Gray, Ill. Ind. Zool. ii. pl. 45, fig. 3 (1834).
Perdicula rubicola, Hodgs. Beng. Sport. Mag. ix. p. 344 (1837).
Perdicula asiatica, Gould, B. As. vii. pl. 4 (1863); Hume and Marshall, Game Birds of India, ii. p. 109, pl. (1879); Oates, ed. Hume's Nests and Eggs Ind. B. iii. p. 440 (1890); Ogilvie-Grant, Cat. Birds Brit. Mus. xxii. p. 198 (1893).
Perdicula argoondah, Gould (*nec* Sykes), B. As. vii. pl. 5 (1863).

Adult Male.—General colour above brown, with pale buff shaft-stripes on the back, and black bars and blotches on the scapulars and wing-coverts; forehead, eyebrow-stripes, and throat rufous-chestnut, with whitish edges; under-parts white, with regular black cross-bars; *inner webs of primary flight-feathers not* barred with rufous-buff. Total length, 6·4 inches; wing, 3·3; tail, 1·5; tarsus, 0·95.

Adult Female.—Has the throat rufous-chestnut like the male, but differs in having no buff shaft-streaks on the upper-parts, and the under-parts uniform vinaceous-buff.

Range.—India and Ceylon.

Habits.—According to Mr. Hume, "moderately thick forests and jungles, hills, ravines, and broken ground, not too deficient in cover, and rich cultivation, if not in too damp and undrained situations, from near the sea-level to an elevation of four to five thousand feet, are the ordinary resorts of the Jungle Bush-Quail. Very considerable differences in rainfall affect them but little, provided the ground is hilly, raviny, or well drained, and cover sufficient, and they are abundant, as on the Western Ghats, where the rainfall is over 100 inches, and on scrub-clad hills in Rajputana, where it certainly falls short of 20 inches."

Tickell says: "They prefer stony, gravelly places, amongst thorny bushes, such as the jujube or bér, or tracts of stunted Sâl, Assun, and Polâs (or Dhâk), congregating in coveys of eight to a dozen under thickets, whence of an evening they emerge into adjacent fields, meadows, and clumps of grass to feed. They lie very close, suffering themselves to be almost trodden upon, and then rise at once out of some small bush, with a piping whistle, and such a sudden start and whir, instantly flying off to all parts of the compass—including sometimes a close shave of the sportsman's countenance,—that a more difficult bird to hit could nowhere be found, especially as their flight is prodigiously rapid, and directed so as barely to skim the upper twigs of the bushes. They do not go far, but, when once down, are hardly ever flushed again till they have reunited. This they lose as little time as possible in doing, running like mice through the herbage to some central spot, where the oldest cock bird of the covey is piping all hands together. Although so gregarious and sociable, these birds are very quarrelsome, and their extreme pugnacity leads to their easy capture. . . . . Bush-Quails are not often caught by hawking, as the Uriyas do not care to trust their trained sparrow-hawks (shickras and besras) so much amongst the jungle. For the table they are hard and tasteless, and they are valued by the natives chiefly for their fighting qualities, which do not appear to degenerate even after long confinement."

Nest.—Neatly made of roots and fine grass, in a depression in the ground sheltered by a bush or tuft of grass.

Eggs.—Slightly pointed ovals, varying in colour from creamy to brownish-white; five to seven in number. Average measurements, 1·0 by 0·85 inch.

II. THE ROCK BUSH-QUAIL. PERDICULA ARGOONDAH.

Coturnix argoondah, Sykes, P. Z. S. 1832, p. 153; id. Trans. Zool. Soc. ii. p. 17, pl. ii. (1841).

Perdicula argoondah, Hume and Marshall, Game Birds of India, ii. p. 117, pl. (1879); Oates, ed. Hume's Nests and Eggs Ind. B. iii. p. 441 (1890); Ogilvie-Grant, Cat. B. Brit. Mus. xxii. p. 200 (1893).

Adult Male.—Like the male of *P. asiatica*, but the upper-parts are barred with buff and black or grey; the rufous on the head and throat is *dull brick-colour*, not bordered with white; and the quills are barred on *the inner* as well as the outer webs with rufous-buff. Total length, 6·5 inches; wing, 3·3; tail, 1·7; tarsus, 1.

Adult Female.—Has the *throat white, tinged with vinaceous*; the upper-parts vinaceous-brown and under-parts dull vinaceous, a few faint buff and dusky markings on the former and the middle of the belly whitish-buff.

Range.—India; ? Ceylon (*Layard*). Mauritius [*introduced*].

Habits.—This species appears to have nearly the same wide and irregular distribution throughout the Peninsula of India as its near ally, the Jungle Bush-Quail, but affects very different localities, the two forms being apparently complementary to one another. The Rock Bush-Quail prefers the dry rocky plains or low hillocks, thinly covered with scattered thorn-bushes, and barren sparsely-cultivated districts; and though both species may occasionally be met with on the same stubble where their ranges meet, it may be generally stated that, where one is found, the other does not occur. The habits of the

Rock Bush-Quail are perfectly similar to those of its ally, and, like it, the birds generally feed in company, even in the breeding-season, when "newly-hatched birds may frequently be seen running amongst half-a-dozen old ones" (*Aitken*).

Nest.—Placed under a tussock of grass, and neatly made of dry grass placed in a shallow, saucer-shaped, depression in the ground.

Eggs.—Five to seven in number, generally rather pointed towards the small end; uniform glossy white, slightly tinged with brownish-buff. Average measurements, 1·02 by 0·84 inch.

THE PAINTED BUSH-QUAILS. GENUS MICROPERDIX.

Microperdix, Gould, B. As. vii. pl. iii. (1862).

Type, *M. erythrorhyncha* (Sykes).

Very similar in size and general appearance to the last genus, but the tail is composed of only ten feathers, and the first primary flight-feather is about equal to the tenth, the sixth being slightly the longest. The *tarsi* in the males are *without* any trace of a spur.

Only three very small species are known.

I. THE PAINTED BUSH-QUAIL. MICROPERDIX ERYTHRORHYNCHA.

Coturnix erythrorhyncha, Sykes, P. Z. S. 1832, p. 153; J. E. Gray, Ill. Ind. Zool. ii. pl. 44, fig. 2 (1834); Sykes, Trans. Zool. Soc. ii. p. 16, pl. 1 (1841).

Perdicula erythrorhyncha, Auctorum, *passim*.

Microperdix erythrorhyncha, Gould, B. As. vii. pl. 3 (1862); Hume and Marshall, Game Birds of India, ii. p. 123, pl. (1879); Oates, ed. Hume's Nests and Eggs Ind. B. iii. p. 442 (1890); Ogilvie-Grant, Cat. B. Brit. Mus. xxii. p. 203 (1893).

Adult Male.—General colour above earthy-brown, with rounded black spots, blotched, especially on the wings, with black; the latter being also marked with whitish shaft-streaks and buff cross-bars; top of the head, except the middle of the crown, black, with a *narrow* well-defined white band between the eyes, continued backwards on each side of the head and forming a U-shaped white mark; throat white; chest greyish-brown, with a rufous wash, and shading gradually into rufous-chestnut on the rest of the under-parts; the sides and flanks with rather large black white-edged spots. Total length, 7 inches; wing, 3·4; tail, 1·9; tarsus, 1.

Adult Female.—Like the male, but the black on the head and the white throat are replaced by dull rufous-chestnut.

Range.—South-western hills of the Peninsula of India, extending from Bombay to the Cardamum Hills in Travancore.

Habits.—Davison says: "The Painted Bush-Quail is very abundant on the Nilgiris and their slopes, and is not uncommon in the Wynád. They always occur in bevies, numbering eight to twelve birds. They of course avoid the inner depths of the jungles, but are found on the outskirts, especially where there is good dense cover, such as the common brake-fern; but their favourite resort is rather rocky ground, interspersed with bushes and dense clumps of fern and high grass, especially when such places abut on or are near cultivation, or any road along which cattle, carrying grain, habitually pass.

"About the station of Ootacamund they are, even to this day, not uncommon; and in the grounds of almost all the outlying houses, where these are tolerably wooded, one or more coveys are sure to be found. In the mornings and evenings they are very fond of coming out into the open, and I have met with a dozen or more coveys on the road in a morning's ride between Coonoor and Ooty. They are tame little birds and will seldom rise when met with on a road, unless hard pressed or suddenly surprised; they content themselves with running on ahead, occasionally stopping to pick up a grain or an insect, until they think they are being too closely followed, when they quietly slip out of sight into the first bit of cover they come to.

"When retreating they keep uttering a very rapidly and continually repeated note, in a very low tone, hardly to be heard unless when one is quite close to them.

"When flushed they do, as a rule, rise, as Jerdon says, all together, usually scattering in different directions, but this is by no means invariably the case, and sometimes, even before a dog, they will rise singly, or in couples, several minutes often intervening between the rise of the first and last birds. . . .

"When a covey has been flushed and scattered, one bird commences after a few minutes calling in a very low tone, another immediately taking it up, then another, and so on. They then begin cautiously to reunite, uttering all the time their low note of alarm, moving very slowly, with continual halts while in cover, but dashing rapidly across any open space they may have to cross."

Nest.—Like that of the Rock Bush-Quail.

Eggs.—Ten or more; rather long ovals. Uniform glossy pale brownish-buff. Average measurements, 1·22 by 0·91 inch.

II. BLEWITT'S PAINTED BUSH-QUAIL. MICROPERDIX BLEWITTI.

Microperdix blewitti, Hume, Str. F. ii. p. 512 (1874); id. and Marshall, Game Birds of India, ii. p. 130, pl. (1879); Ogilvie-Grant, Cat. B. Brit. Mus. xxii. p. 204 (1893).

Adult Male.—Closely resembles the male of *M. erythrorhyncha*, but is smaller and distinguished by having the white band between the eyes much wider and the black forehead much narrower, the chest greyer, and the rest of the under-parts paler. Total length, 6·8 inches; wing, 3·1; tail, 1·6; tarsus, 0·95.

Adult Female.—Like the female of *M. erythrorhyncha*, but paler.

Range.—Central Provinces of India.

III. THE MANIPUR PAINTED BUSH-QUAIL. MICROPERDIX MANIPURENSIS.

Perdicula manipurensis, Hume, Str. F. ix. p. 467 (1880).
Microperdix manipurensis, Ogilvie-Grant, Cat. B. Brit. Mus. xxii. p. 204 (1893).

(*Plate XIII.*)

Adult Male.—Above dark grey, barred (and blotched on the wings) with black ; forehead, eyebrow-stripes, and throat *dark chestnut;* neck and chest grey, shading into *tawny* on the rest of the under-parts, which have a black shaft-stripe and wide black cross-bar on each feather. Total length, 6·5 inches ; wing, 3·4 ; tail, 2 ; tarsus, 1·05.

Adult Female.—Differs from the male in having *no* chestnut on the head or throat, the latter being *whitish*, and the breast and belly are buff.

Range.—Sikhim and South-eastern Manipur Hills.

This extremely handsome little species was discovered by Mr. A. O. Hume during his expedition to Manipur, and was only met with in one place, in a patch of thick elephant-grass jungle, where eleven adult and immature specimens were obtained. A single bird was subsequently shot in the same district, and there is also a skin of this species in the British Museum which is said to have been obtained in Sikhim. No doubt the bird occurs in the intermediate districts in suitable localities and will be found by future collectors, but owing to its skulking habits and small size, it has hitherto been overlooked and the only specimens so far known are those mentioned above. It appears to live entirely in the almost impenetrable patches of elephant-grass, only venturing into the more open spaces in the early morning when feeding, and never rises till very hard pressed, preferring, if possible, to escape by running. The specimens obtained by Mr. Hume (numbering two coveys of six and five birds respectively, all of which were shot) were first seen in an open glade in a patch of elephant-grass about two miles square, and with the aid of about a hundred beaters were eventually obtained after two days' arduous work. After reading his account of how these specimens were obtained (in the volume of "Stray Feathers" cited above), some idea may be formed of the labour and expense entailed in forming the magnificent "Hume collection" of Indian birds which, thanks to the generosity of that great ornithologist, now forms part of the National Collection.

THE TREE-PARTRIDGES. GENUS ARBORICOLA.

Arboricola, Hodgs. in Gray's Zool. Misc. p. 85 (1844).

Type, *A. torqueola* (Valenc.).

Tail less than half the length of the wing, composed of fourteen short, somewhat rounded feathers, the middle pair being rather the longest.

First flight-feather intermediate in length between the eighth and tenth; fourth and fifth slightly the longest.

Throat and fore-part of the neck often thinly covered with feathers or nearly naked.

Nails *long and nearly straight*.

A supra-orbital chain of bones[*] (fig. 1).

Sexes similar in plumage or very nearly so, with the exception of *A. torqueola*.

A concealed patch of downy feathers on the sides of the body under the wings, *grey*.

1. COMMON TREE-PARTRIDGE. ARBORICOLA TORQUEOLA.

Perdix torqueola, Valenc. Dict. Sci. Nat. xxxviii. p. 435 (1825).

[*] Mr. W. T. Blanford has called my attention to the fact that two specimens of *A. chloropus* in the British Museum bear MS. notes by Mr. Wood-Mason stating that the supra-orbital chain of bones peculiar to all the other species of *Arboricola* is *absent*. *A. chloropus* and *A. charltoni* belong to

Perdix megapodia, Temm. Pl. Col. v. pls. 35 and 36 [Nos. 462, 463] (1828).
Perdix olivacea, J. E. Gray, Ill. Ind. Zool. i. pl. 57 (1830-32).
Arboricola torqueola, Hume and Marshall, Game Birds of India, ii. p. 69, pl. (1879); Ogilvie-Grant, Cat. B. Brit. Mus. xxii. p. 207 (1893).

Adult Male.—Crown *bright chestnut*; back olive-brown, *barred with black*; wing widely margined with chestnut, and blotched with black on the coverts; sides of the face and throat black,

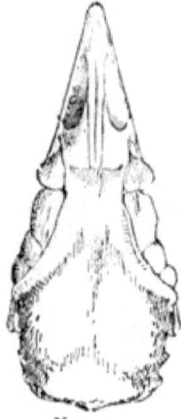
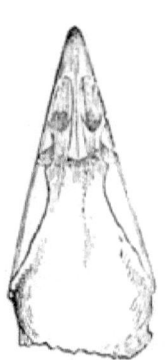

FIG. 1.   FIG. 2.

the feathers more or less edged with white; fore-neck white; chest grey, shading into white on the under-parts; flank-feathers *grey, edged with chestnut and with a white central spot*. Total length, 11·8 inches; wing, 6; tail, 2·7; tarsus, 1·7.

Adult Female.—Differs chiefly from the male in having the crown *brown* with black shaft-stripes; the sides of the face, throat, and neck *rust-coloured, spotted with black*; and the white central spots on the flank-feathers much larger.

rather a distinct group of *Arboricola*, and on examination we find that the supra-orbital chain of bones is wanting *in both* species (fig. 2). This being the case, Mr. Blanford proposes, very rightly as we think, to place these two species in a different genus, for which the name *Tropicoperdix*, Blyth, has already been proposed.

Range.—Outer ranges of the Himalayas, from 5,000 to 14,000 feet, and extending from the eastern borders of Chamba to Sikhim, and southwards to the Manipur Hills.*

Wilson gives the following account of this species:—"This handsome little Partridge inhabits the forests and jungles, and is never found in open spots or cultivated fields. It is most numerous on the lower ranges in the wooded ravines and hill-sides, from the summit to near the base, but does not occur at the foot of the hills or low down in the valleys. It is not so common in the interior, but is met with to an elevation of about 9,000 feet. It is rather solitary in its habits, generally found in pairs; but occasionally, in autumn and winter, five or six will collect together and keep about one spot.

"It is a quiet, unsuspicious bird; when alarmed it utters a soft whistle, and generally creeps away through the underwood, if not closely pressed, in preference to rising. Its flight is rapid, oftener across the hill than downwards, and seldom very far; in general, not more than eighty or one hundred yards. Its food being very similar, it is met with in the same places as the Koklass Pheasant, and both are often found together. Indeed, in winter, in some of the forests of the interior, Tragopans, Moonal, Koklass, and Kalij Pheasants, and the Hill Partridge, are at all times all found within a compass of fifty or sixty yards.

"It feeds on leaves, roots, maggots, seeds, and berries; in confinement it will eat grain; in a large cage or enclosure its motions are very lively, and it runs about with great sprightliness from one part to another. It occasionally mounts into the trees, but not so often as a forest bird might be expected to do.

"In the forests of the interior, in spring, it is often heard calling at all hours of the day. The call is a single loud soft whistle, and may be easily imitated so as to entice the bird

---

* I have seen specimens obtained by Col. Godwin-Austen in this locality.

quite close. At other seasons it is never heard to call, except when disturbed."

Hodgson says that "they constantly perch. At the top of Pulchook I flushed a covey of eight or ten, which flew widely scattered, all alighting on the highest trees."

Eggs.—Said (no doubt correctly) to be pure white, and six to eight in number.

II. THE BLACK-THROATED TREE-PARTRIDGE. ARBORICOLA ATRIGULARIS.

Arboricola atrogularis, Blyth, J. As. Soc. Beng. xviii. p. 819 (1849); Hume and Marshall, Game Birds of India, ii. p. 79, pl. (1879); Oates, ed. Hume's Nests and Eggs Ind. B. iii. p. 439 (1890); Ogilvie-Grant, Cat. B. Brit. Mus. xxii. p. 209 (1893).

Adult Male and Female.—Above olive-brown, barred with black on the back, and much the same as in *A. torqueola*; throat black; feathers of fore-neck black, edged with white; chest grey, shading into white on the middle of the upper-parts; flank-feathers grey, edged with olive-brown, and with an oval white spot near the end of the shafts. Total length of *male*, 10·0 inches; wing, 5·6; tail, 2·2; tarsus, 1·7. *Female* rather smaller.

Range.—North-eastern India, extending from Eastern Assam to Chittagong and Manipur, and eastwards to the Kachin Hills east of Bhâmo.

Habits.—Mr. J. R. Cripps found this species common in Sylhet, frequenting the hillocky ground covered with dense forest. He says:—"They are very fond of feeding about the banks of the small rivulets that meander among the hillocks which are scattered about the district. They feed on insects, for which they scratch amongst the decaying leaves that carpet the ground, seeds and berries of various kinds, and on young shoots.

"Though greatly affecting dense forests, these birds are also partial to bamboo-jungle. . . . .

"Their call, which is often heard, especially towards dusk, is a rolling whistle, *whew, whew,* repeated many times, and winding up with a sharper and more quickly uttered *whew.* The sound is very easily imitated, and the birds are easily enticed to approach one by the imitation, and this is the way in which natives usually secure them.

"This species certainly perches at times, for I have seen one fly down from a small tree."

Nest.—A mere lining of leaves and twigs placed in a slight depression at the foot of a large tree (*Cripps*).

Eggs.—Four in number; pure white, broad ovals, rather pointed towards the small end; shell fine, rather glossy. Average measurements, 1·38 by 1·12 inch.

III. THE HAINAN TREE-PARTRIDGE. ARBORICOLA ARDENS.

Arboricola ardens, Styan, Ibis, 1893, pp. 56, 436, pl. xii.; Ogilvie-Grant, Cat. B. Brit. Mus. xxii. p. 210 (1893).

Adult Male.—In general plumage like *A. atrigularis,* but with an *orange-red patch* of feathers on the fore-part of the neck and middle of the chest, the sides of the neck and under-parts being faintly washed with the same colour. Total length, 8 inches; wing, 4·8; tail, 1·75; tarsus, 1·3.

Range.—Mountains of Hainan.

This remarkable species is at present only known from a single skin, the property of Herr Schmacker.

IV. THE FORMOSAN TREE-PARTRIDGE. ARBORICOLA CRUDIGULARIS.

Oreoperdix crudigularis, Swinhoe, Ibis, 1864, p. 426.
Arboricola crudigularis, Blyth; Ogilvie-Grant, Cat. B. Brit. Mus. xxii. p. 211 (1893).

Adult Male and Female.—Very similar to *A. atrigularis,* but the barring on the back is wider and darker; the chin, upper-part of the throat and fore-neck are white, and the lower part of the throat black. Total length, 9·5 inches; wing, 5·5; tail, 2·2; tarsus, 1·55.

Range.—Mountains of the interior of Formosa.

V. THE ARACAN TREE-PARTRIDGE. ARBORICOLA INTERMEDIA.

Arboricola intermedia, Blyth, J. As. Soc. Beng. xxiv. p. 277 (1856); Hume and Marshall, Game Birds of India, ii. p. 85, pl. (1879); Oates, B. Burmah, ii. p. 327 (1883); and ed. Hume's Nests and Eggs Ind. B. iii. p. 440 (1890); Ogilvie-Grant, Cat. B. Brit. Mus. xxii. p. 211 (1893).

Adult Male and Female.—Upper-parts olive-brown *without* black bars on the back, though sometimes the feathers have dusky margins; otherwise the plumage is much like that of the *female* of *A. torqueola*, but the chin and throat are *black*, and the fore-neck bright rufous.

Range.—North-eastern India, extending from the Garo and Nága Hills to the borders of Arakan and Pegu and eastwards to Bhâmo.

VI. THE RUFOUS-THROATED TREE-PARTRIDGE. ARBORICOLA RUFIGULARIS.

Arboricola rufogularis, Blyth, J. As. Soc. Beng. xviii. p. 819 (1849); Hume and Marshall, Game Birds of India, ii. p. 75, pl. (1879); Oates, ed. Hume's Nests and Eggs Ind. B. iii. p. 439 (1890); Ogilvie-Grant, Cat. B. Brit. Mus. xxii. p. 212 (1893).

Adult Male and Female.—Resemble the sexes of *A. intermedia*, but the feathers of the chin and throat are rust-red spotted with black, as in the *female of A. torqueola;* from the latter they are easily distinguished by the *absence* of black bars on the back. Total length, 11 inches; wing, 5·6; tail, 2·3; tarsus, 1·7.

Range.—Lower outer ranges of the Himalayas from Kumaon

to at least as far east as the Darrung district, north of the Daphla Hills. It is also common on the higher ranges of Tenasserim, on Mooleyit, but has not been met with in the intervening countries.

Habits.—The Rufous-throated Tree-Partridge, which, in general appearance, closely resembles the female of the Common Tree-Partridge (*A. torqueola*) but is at once distinguished by the absence of black bars on the back, inhabits a much lower range than that species, not exceeding an elevation of about 6,000 feet in the summer, while in cold weather it may be met with nearly at the bottom of the valleys. It is also more often met with in coveys than the common species, but its general habits are otherwise perfectly similar, though apparently it is more given to perching on trees, when flushed by a dog or otherwise. Davison, who collected many on the higher slopes of Mooleyit in Tenasserim, tells us that he has shot three or four when thus perched, before the others attempted to move, and that these birds sometimes settled in trees within a few feet of him, being apparently far tamer than the Himalayan examples of this species.

Nest and Eggs.—Very similar to those of *A. atrigularis*. An egg measures 1·5 by 1·2 inch.

VII. SONNERAT'S TREE-PARTRIDGE. ARBORICOLA GINGICA.

La Perdrix de Gingi, Sonnerat, Voy. Ind. Orient. ii. p. 167 (1782).

Tetrao gingicus, Gmel. S. N. i. pt. ii. p. 760 (1788).

Arboricola gingica, Blyth, Ibis, 1870, p. 174; Ogilvie-Grant, Ibis, 1892, 395, pl. ix.; id. Cat. B. Brit. Mus. xxii. p. 213 (1893).

Adult.—General colouring above similar to that of *A. rufigularis*, but easily distinguished by having *a triangular black patch on the base of the fore-neck, succeeded by a narrower white, and a wider band of deep maroon.*

PLATE XIV.

MANDELLI'S TREE-PARTRIDGE.

Although this species has been known to science for more than a century, its habitat still remains unknown; and, so far as I am aware, the only specimen at present known is that in the Leyden Museum. It has been suggested that the Philippine Islands might probably prove to be the home of this bird, but although most of them have now been visited by various naturalists, we are still no nearer the solution of the mystery.

VIII. MANDELLI'S TREE-PARTRIDGE.   ARBORICOLA MANDELLII.

Arborophila mandellii, Hume, Str. F. ii. p. 449 (1874); iii. p. 262, pl. 1 (1875).

Arboricola mandellii, Hume and Marshall, Game Birds of India, ii. p. 83, pl. (1879); Ogilvie-Grant, Cat. B. Brit. Mus. xxii. p. 214 (1893).

(*Plate XIV.*)

Adult Male and Female.—Crown dark chestnut, shading into brown on the nape; sides of the neck and cheeks rust-red, spotted with black; feathers of the back olive-brown, margined but not barred with dusky; eyebrow-stripes grey; throat and fore-neck rust-red, *divided from the rich chestnut chest by a white and a black band;* breast and belly grey. Total length, 11 inches; wing, 5·4; tail, 2·2; tarsus, 1·6.

Nothing is known of the habits or nidification of this remarkable species, the only specimens as yet obtained having been collected by the late Mr. Mandelli's hunters in the damp dense jungles of the Bhotan Doars and Native Sikhim. The whole of that ornithologist's splendid collection, having been purchased at his death by Mr. A. O. Hume, now forms part of the Hume collection in the British Museum.

IX. THE JAVAN TREE-PARTRIDGE.   ARBORICOLA JAVANICA.

Javan Partridge, Brown, Ill. Zool. p. 40, pl. 17 (1776).
Tetrao javanicus, Gmel. S. N. i. pt. 11, p. 761 (1788).

Arboricola javanica, Ogilvie-Grant, Cat. B. Brit. Mus. xxii. p. 214 (1893).

Adult Male and Female.—Crown reddish-brown, shading into rusty on the forehead and bordered by a black band, which encircles the eye, and is continued down the middle of the rust-red nape, and joins a second black band surrounding the base of the neck; *upper-parts grey*, barred with black; cheeks and throat rust-coloured; chest grey; *rest of under-parts and flanks chestnut*. Total length, 11 inches; wing, 5·6; tail, 2·2; tarsus, 1·8.

Range.—Mountains of Java.

X. THE RED-BILLED TREE-PARTRIDGE. ARBORICOLA RUBRIROSTRIS.

Peloperdix rubrirostris, Salvadori, Ann. Mus. Civ. Genov. xiv. p. 251 (1879); Snelleman, in Veth's Midden-Sumatra, iv. p. 46, pl. iii. (1887).

Arboricola rubrirostris, Ogilvie-Grant, Cat. B. Brit. Mus. xxii. p. 215 (1893).

Adult Male.—Head, throat, and neck black, with a few white spots on the sides of the crown and throat, and a small patch of the same colour on the chin; upper-parts reddish olive-brown, barred with black; chest brown; upper breast and sides of the belly white, with a large black spot on each feather; middle of belly white; *flanks black, barred with white*. Total length, 9·6 inches; wing, 5·2; tail, 1·7; tarsus, 1·75.

Adult Female.—Has rather more white on the lores and chin and is somewhat smaller.

Range.—Mountains of Sumatra.

This is a very rare bird in collections, very few specimens having as yet been obtained, but its habits, so far as they are known, appear to be similar to those of its allies. The coloration of the under-parts is peculiar, and very different from that of any of the other species of this genus.

XI. THE BROWN-BREASTED TREE-PARTRIDGE. ARBORICOLA BRUNNEIPECTUS.

Arboricola brunneopectus, Tickell; Blyth, J. As. Soc. Beng. xxiv. p. 276 (1855); Hume and Marshall, Game Birds of Ind. ii. p. 87, pl. (1879); Ogilvie-Grant, Cat. B. Brit. Mus. xxii. p. 216 (1893).

Adult Male and Female.—Crown brownish-black; forehead, wide eyebrow-stripes, cheeks, and throat buff; a black band commencing at the gape, surrounds the eye, and ends in a black patch on the side of the neck; upper-parts olive-brown, barred with black; wings marked with pale olive and chestnut, blotched with black; sides and front of neck spotted with black; chest brownish-ochre, shading into whitish on the belly; flank-feathers with a *large white spot near the extremity, partially or wholly bordered with black*. Total length, 11 inches; wing, 5·6; tail, 2·5; tarsus, 1·7.

Range.—Evergreen forests of Burma and North Tenasserim; extending from the Karen and Tonghoo Hills through Eastern Pegu as far south as Tavoy, Tenasserim.

Habits.—This species is met with from nearly sea-level to an elevation of about 4,500 feet. Mr. Darling, who had many opportunities of studying its habits in the vicinity of Thoungyah, usually found it between the months of September and November in coveys of from three to ten or even more birds, "but," he says, "owing to their shyness and dead-leaf colour, they were difficult to secure. They feed amongst the dead leaves on seeds, insects, and small shells, and are very restless, giving a scratch here, a short run and another scratch there, and so on, uttering a soft cooing whistle all the time. When disturbed by a man, they always disappeared into the dense undergrowths; but a dog always sent them flying into some small tree, whence they would at once begin calling to one another, whistling first low and soft, and going up higher and shriller, till the call was taken up by another bird. I often got quite close to them, but the instant

I was seen, away they ran helter-skelter in all directions, and I could only now and then catch a glimpse of the little fellows scuttling through the bushes. Of course they are entirely a forest bird, though they may be seen just at the outskirts of a wood."

XII. TREACHER'S TREE-PARTRIDGE. ARBORICOLA HYPERYTHRA.

Bambusicola hyperythra, Sharpe, Ibis, 1879, p. 266; Gould, B. Asia, vi. pl. 71 (1879).

Arboricola hyperythra, Ogilvie-Grant, Cat. B. Brit. Mus. xxii. p. 217 (1893).

Adult.—Differs from *A. brunneipectus* in having the crown jet black; the broad eyebrow-stripes, ear-coverts, and cheeks *ashy-grey*; and the chin, throat, and under-parts reddish-chestnut. The flank-feathers have the same peculiar black and white marking as in the last species. Total length, 10·5 inches; wing, 5·2; tail, 2·2; tarsus, 1·6.

Range.—North-west Borneo; mountains above the Lawas River.

Since the present species was described in 1879 by Dr. Bowdler Sharpe, from a single specimen obtained by Mr. Treacher in the above locality, no more examples have been obtained, and the type in the Oxford Museum remains unique. It is just possible that a larger series of specimens than we have at present had the opportunity of examining, may show that this species and the following (*A. erythrophrys*) are stages of plumage of the same bird (*see* Sharpe, Ibis, 1894, p. 539), but we do not believe this to be the case, for in none of the specimens of the latter species that we have examined—and we have had a good series before us—is the grey eyebrow-stripe (apparently a sign of immaturity) at all marked, and it is never found in adult specimens with the crown black.

XIII. WHITEHEAD'S TREE-PARTRIDGE. ARBORICOLA ERYTHROPHRYS.

Bambusicola erythrophrys, Sharpe, Ibis, 1890, pp. 139, 284, 288, 289, pl. iv.

Arboricola erythrophrys, Ogilvie-Grant, Cat. B. Brit. Mus. xxii p. 218 (1893).

Adult Male.—Closely allied to the last species, *A. hyperythra*, but distinguished by having the eyebrow-stripes *black*, like the crown, in the fully adult birds, *rust-red* in less mature birds, and greyish in the young (which have the crown *brown*, only slightly spotted with black, and the throat whitish-buff); chin and throat black in the fully adult, rust-coloured in younger birds. Total length, 10 inches; wing, 5·8; tail, 1·8; tarsus, 1·85.

Adult Female.—Appears to differ from the male in never getting the black on the throat.

Range.—North Borneo; the dense bamboo-jungles of Mount Kina Balu, from 2,000 to 4,000 feet.

XIV. HORSFIELD'S TREE-PARTRIDGE. ARBORICOLA ORIENTALIS.

Perdix orientalis, Horsfield, Trans. Linn. Soc. xiii. p. 184 (1822).

Perdix personata, Horsfield, Zool. Res. Java, pl. 61 (1824).

Arboricola orientalis, Ogilvie-Grant, Cat. B. Brit. Mus. xxii. p. 218 (1893).

Adult.—Crown and nape blackish-brown; eyebrow-stripe, sides of the head, and throat white; upper-parts dark brown, fringed with blackish; wing mixed with olive-brown and orange-red, and blotched with black; a dark brown band along each side of the head enclosing the eye; chest and breast brownish-grey; belly whitish; flanks grey, with wide irregular bars of black and white. Total length, 11 inches; wing, 5·9; tail, 2·5; tarsus, 1·8.

Range.—Mountains of East Java; forests of 3,000 feet.

Nothing is known of the habits of this rare bird, which, so far as we are aware, is only known from the unique type obtained by Horsfield in the province of Blambangan, East Java.

XV. THE SUMATRAN TREE-PARTRIDGE. ARBORICOLA SUMATRANA.

Arborophila sumatrana, Ogilvie-Grant, Ann. Mag. N. H. (6), viii. p. 297 (1891).

Arboricola sumatrana, Ogilvie-Grant, Cat. B. Brit. Mus. xxii. p. 219 (1893).

Adult.—Differs from *A. orientalis* in having the crown and upper-parts golden-brown, fringed and strongly barred with black; no white eyebrow-stripe; and the flank-feathers *with broad regular three-fold bands of black, white and black* at the extremity.

Range.—Mountains of Central Sumatra, at about 3,000 feet.

The type-specimen has been in the Museum for many years, but its origin is unknown. Dr. H. O. Forbes procured some specimens during his travels in Sumatra, but no one else seems to have met with the species.

THE WOOD-PARTRIDGES. GENUS TROPICOPERDIX.

Tropicoperdix, Blyth, J. As. Soc. Beng. xxviii. p. 415 (1859).

Type, *T. chloropus*, Blyth.

Characters the same as those given for *Arboricola*, but the peculiar supra-orbital chain of bones is *wanting*[*] and the concealed patch of downy feathers on each side of the body under the wing is *pure white*.

I. THE GREEN-LEGGED WOOD-PARTRIDGE. TROPICOPERDIX CHLOROPUS.

Tropicoperdix chloropus (Tickell), Blyth, J. As. Soc. Beng. xxviii. p. 415 (1859).

---

[*] See fig. 2, p. 161, and footnote, p. 160.

Arboricola chloropus, Tickell, J. As. Soc. Beng. xxviii. p. 453 (1859); Hume and Marshall, Game Birds of India, ii. p. 91, pl. (1879); Ogilvie-Grant, Cat. B. Brit. Mus. xxii. p. 219 (1893).

Adult Male and Female.—Crown and upper-parts warm brown, barred and marked with black; wing-coverts mixed with buff and rufous; eyebrow-stripes black and white, and bordering the crown and sides of the nape; sides of face and throat white, the neck rust-colour, all these parts dotted with black; chest brown, barred with black; middle of the breast rust-colour, shading into whitish on the belly; flanks buff, *irregularly* barred and marked with black.

Range.—From the bases of the hills north of Tonghoo and the Eastern Pegu Hills, and extending as far south as Tavoy in Tenasserim. Also recorded from Cochin China.

Habits.—This species appears in Tenasserim to be confined to the lower forests and jungles that skirt the bases of the hills, generally avoiding the more dense hill-forests of the higher elevations, where *Arboricola brunneipectus* is met with. Davison found it most abundant in the thin tree-jungle, generally in pairs or small parties which were seen gliding about on the ground amongst the dense brushwood, and scratching amongst the dead leaves in search of insects and seeds. He says that the note is a soft double whistle, chiefly uttered in the morning and evening, and that without dogs they are hard to procure, being loath to take wing and preferring to run and squat under cover. When flushed by a dog they rise at once, and after flying a short distance drop to the ground, never perching on trees as is often the habit of *Arboricola rufigularis* and its allies.

II. CHARLTON'S WOOD-PARTRIDGE. TROPICOPERDIX CHARLTONI.

Perdix charltoni, Eyton, Ann. and Mag. N. H. xvi. p. 230 (1845).

Tropicoperdix charltoni, Blyth, J. As. Soc. Beng. xxviii. p. 415 (1859).

Arboricola charltoni, Hume and Marshall, Game Birds of India, ii. p. 93, pl. (1879); Ogilvie-Grant, Cat. B. Brit. Mus. xxii. p. 221 (1893).

Adult Male and Female.—Differ chiefly from the sexes of *T. chloropus* in having the upper-parts finely mottled with black and marked with irregular buff spots and bars; a black band on the sides and base of the neck; the upper half of the chest deep chestnut; and the flank-feathers *regularly* barred with black and buff. Total length, 11 inches; wing, 6·3; tail, 2·5; tarsus, 1·6.

Range.—The hill-jungles of the Malay Peninsula from Pinang southwards, Sumatra, and North Borneo. Reported also from Bangkok in Siam and the South Tenasserim Hills.

THE CRIMSON-HEADED WOOD-PARTRIDGES. GENUS HÆMATORTYX.

Hæmatortyx, Sharpe, Ibis, 1879, p. 266.

Type, *H. sanguiniceps*, Sharpe.

Tail short and rounded, composed of twelve feathers, and less than half the length of the wing.

First primary flight-feather equal to the tenth, fifth slightly the longest. Feet armed with *three* pairs of spurs in the *male* (none in the female); hind-toe with a small but well-developed claw.

Sexes different.

The only species known is

1. THE CRIMSON-HEADED WOOD-PARTRIDGE. HÆMATORTYX SANGUINICEPS.

Hæmatortyx sanguiniceps, Sharpe, Ibis, 1879, p. 266; Gould, B. Asia, vi. pl. 70 (1879) [♀]; Ogilvie-Grant, Cat. B. Brit. Mus. xxii. pp. 222 and 560 (1893); id. Ibis, 1894, p. 374, pl. x. [♂].

Adult Male.—General colour blackish-brown; crown and nape dull deep crimson; fore-neck, chest, and longer under tail-coverts deep brilliant crimson; cheeks and throat paler crimson. Total length, 10·5 inches; wing, 6·6; tail, 2·8; tarsus, 2·3.

Adult Female.—Differs from the *male* in having the throat pale rufous, washed with crimson, and the fore-part of the neck and chest reddish-chestnut. Smaller. Total length, 10·5 inches; wing, 6; tail, 2·8; tarsus, 1·9.

Range.—Mountain forests and jungles of North Borneo.

This splendid Wood-Partridge was first described by Dr. Bowdler Sharpe, from a single female specimen obtained by Mr. W. H. Treacher near the Lawas River and now in the Oxford Museum. In 1891 a second female was collected by Mr. Hose on the moss-clad summit of Mount Dulit, at an elevation of about 5,000 feet; but it was not until 1893, when Mr. Everett's collectors captured a third example of this rare bird on the eastern slope of Mount Kina Balu, that the *adult male* of this wonderful form of Partridge, with its brilliant crimson chest and treble-spurred legs, became known. The three individuals mentioned are all that have been obtained, and the two latter now form part of the National Collection.

THE FERRUGINOUS WOOD-PARTRIDGES. GENUS CALOPERDIX.

Caloperdix, Blyth, Ibis, 1867, p. 160.

Type, *C. oculea* (Temm.).

Tail rather short and composed of fourteen feathers; less than half the length of the wing.

First primary flight-feather equal to the tenth, the fourth to the sixth feathers being equal and longest.

Tarsi armed in the *male* with one or more pairs of spurs; nail on the hind-toe *rudimentary*.

1. THE FERRUGINOUS WOOD-PARTRIDGE. CALOPERDIX OCULEA.

Perdix oculea, Temm. Pig. et Gall. iii. pp. 408 and 732 (1815).

Caloperdix oculea, Blyth, Ibis, 1867, p. 160; Hume and Marshall, Game Birds of India, ii. p. 101, pl. (1879); Ogilvie-Grant, Cat. B. Brit. Mus. xxii. p. 222 (1893).

Adult Male and Female.—Crown chestnut; nape rufous, *shading into black* on the mantle, each feather of which is marked with *concentric* white lines; feathers of rest of upper-parts black, with a bright rust-coloured sub-marginal band; wings olive-brown, with round black spots; sides of the head and throat rufous-buff shading into clear rufous-chestnut on the rest of the under-parts; sides barred with whitish and black.

Male: Total length, 11·8 inches; wing, 5·7; tail, 2·5; tarsus, 1·8.

Female: Somewhat smaller.

Range.—The dense forests of South Tenasserim and the Malay Peninsula.

Nothing is known of the habits of this species, except that it feeds on insects, seeds, and berries, and all the specimens obtained have been caught in snares.

SUB-SP. *a*. THE SUMATRAN FERRUGINOUS WOOD-PARTRIDGE.
CALOPERDIX SUMATRANA.

Caloperdix sumatrana, Ogilvie-Grant, Bull. Brit. Orn. C. No. ii. p. v. (1892); id. Ibis, 1892, p. 118; id. Cat. B. Brit. Mus. xxii. p. 224 (1893).

Adult Male and Female.—This sub-species only differs from the typical *C. oculea* in having the black feathers of the mantle marked with *irregular transverse bands of pale yellow*, the under-parts less brightly coloured, and the basal half of the breast-feathers mottled and barred with black.

Range.—Forests and jungles of Sumatra and Java.

II. THE BORNEAN FERRUGINOUS WOOD-PARTRIDGE.
CALOPERDIX BORNEENSIS.

Caloperdix borneensis, Ogilvie-Grant, Bull. Brit. Orn. C. No. ii. p. v. (1892); id. Ibis, 1893, p. 117; id. Cat. B. Brit. Mus. xxii. p. 224 (1893).

RED-CRESTED-WOOD-PARTRIDGE.

Adult Male.—Also very closely allied to *C. oculea*, but the mantle is very black and *sharply defined* from the chestnut of the crown and nape, the *concentric white lines* are narrower and more regular than in typical *C. oculea* from the Malay Peninsula, and the throat is darker and more rufous-chestnut.

Range.—Mount Dulit, Sarawak, North Borneo.

The only known example of this extremely handsome species was obtained by Mr. C. Hose in the month of May on the moss-clad summit of Mount Dulit, at an elevation of 5,000 feet.

THE CRESTED WOOD-PARTRIDGES. GENUS ROLLULUS

Rollulus, Bonnat. Tabl. Encycl. Méth. i. Introd. p. xciii. (1790).
Type, *R. roulroul* (Scop.).

A tuft of *long hair-like bristles* on the middle of the forehead.

A long full hairy crest in the *male*.

Tail short, soft, and rounded; composed of twelve feathers; about two-fifths of the length of the wing.

First primary flight-feather equal in length to the tenth, fifth slightly the longest.

Claw on the hind-toe *quite rudimentary*.

A naked patch of skin round the eye. Sexes entirely different in plumage. Only one species is known.

I. THE RED-CRESTED WOOD-PARTRIDGE. ROLLULUS ROULROUL.

Phasianus roulroul, Scop. Del. Flor. et Faun. Insubr. ii. p. 93 (1786).
Phasianus cristatus, Sparrm. Mus. Carls. fasc. iii. pl. 64 (1788).
Tetrao viridis, Gmel. S. N. i. pt. ii. p. 761 (1788) [*female*].
Tetrao porphyrio, Shaw and Nodd. Nat. Misc. iii. pl. 84.
Rollulus roulroul, auctorum, *passim*; Hume and Marshall, Game Birds of India, ii. p. 103, pl. (1879); Ogilvie-Grant, Cat. B. Brit. Mus. xxii. p. 225 (1893).

(*Plate XV.*)

Adult Male.—Head and neck black, except a white band between the eyes, and the long, hairy, maroon-coloured crest which covers the hinder part of the head; upper-parts rich green, glossed with steel-blue; wings dark brown, mixed with buff, except the inner coverts and scapulars, which are maroon, glossed with purplish-blue; under-parts black, glossed with blue. Total length, 10·8 inches: wing, 5·5; tail, 2·3; tarsus, 1.7.

Adult Female.—Head blackish-grey, moderately crested; rest of the plumage bright grass-green, washed with grey on the belly, except the wings, which resemble those of the male, but the inner wing-coverts and scapulars are mostly chestnut, only edged with maroon, and but slightly glossed. Size rather smaller.

Range.—From Southern Tenasserim and Western Siam, south through the Malay Peninsula to Sumatra, Java, Billiton, and Borneo.

Habits.—The only notes on the habits of this bird worth recording are those by the late Mr. W. Davison, who writes: "This species is always found in small parties of six or eight or more, males and females, keeping to the dense forest, and never venturing into the open, living on berries, seeds, tender shoots and leaves, and insects of various sorts. They do not scratch about nearly so much as the *Arboricolas*, and are much quicker and more lively in their movements, much like a Quail, running hither and thither. They rise well before a dog, but it is hard to flush them without. Their note is a soft, mellow, pleasant whistle, which is chiefly heard in the morning, but which they also utter when calling to each other after they have been separated. Like that of *A. rufigularis*, their note is very easily imitated, and they will answer the call readily."

THE BLACK WOOD-PARTRIDGES. GENUS MELANOPERDIX.

Melanoperdix, Jerd. B. Ind. iii. p. 580 (1864).

Type, *M. nigra* (Vig.).

Tail short, soft, and rounded; composed of twelve feathers; more than half the length of the wing.

First primary flight-feather equal in length to the tenth, fifth rather the longest.

No occipital crest in either sex.

Bill unusually stout and thick.

Hind-toe with a *rudimentary* claw. Sexes entirely different in plumage.

Only one species is known.

I. THE BLACK WOOD-PARTRIDGE. MELANOPERDIX NIGRA.

Cryptonyx niger, Vigors, Zool. Journ. iv. p. 349 (1829) [*male*].

Cryptonyx ferrugineus, Vigors, Zool. Journ. iv. p. 349 (1829) [*female*].

Cryptonyx dussumieri, Less. Bélang. Voy. Ind. p. 275, pl. vii. (1834).

Melanoperdix nigra, Jerd. B. Ind. iii. p. 580 (1864); Ogilvie Grant, Cat. B. Brit. Mus. xxii. p. 228 (1893).

Adult Male.—Entire plumage uniform glossy black. Total length, 10·5 inches; wing, 5·5; tail, 3·3; tarsus, 1·8.

Adult Female.—General colour above chestnut, finely mottled with black, and with bars of the same colour on the scapulars; throat and belly whitish; chest dark chestnut; breast and flanks more rufous, the latter barred and mottled with black.

Range.—Southern part of the Malay Peninsula from Province Wellesley southwards, Sumatra, and Borneo.

This species, Mr. C. Hose tells us, is "found in the low country, and does not ascend the mountains" of the Baram district, Sarawak.

Eggs.—Five in number; broad ovals in shape, considerably pointed at the smaller end; white, with the surface slightly rough and chalky. Measurements, 1·65 by 1·3 inch.

THE QUAILS. GENUS COTURNIX.

Coturnix, Bonn. Enc. Méth. Intr. pp. lxxxvii. 216 (1790).

Type, *C. coturnix* (Linn.).

Tail composed of ten or twelve feathers, short, soft, and

hidden by the upper tail-coverts; less than half the length of the wing.

First primary flight-feather about equal to the third, the second being generally slightly the longest: in some instances the first three feathers are sub-equal, or the first may even be a trifle the longest.

Axillary feathers* long and white.

Feet without spurs.   Sexes different in plumage.

This genus may be divided into two sections :

A. Outer web of the primary flight-feathers with irregular bars and marks of buff (species 1 and 2, pp. 180-184).

B. Outer web of the primary flight-feathers uniform brown (species 3 to 6, pp. 185-188).

A. Outer web of the primary flight-feathers with irregular bars and marks of buff.

1. THE MIGRATORY QUAIL.   COTURNIX COTURNIX.

Tetrao coturnix, Linn. S. N. i. p. 278 (1766).
Perdix coturnix, Lath. Ind. Orn. ii. p. 651 (1790).
Coturnix communis, Bonn. Tabl. Encycl. Méth. i. p. 217, pl. 96, fig. 2 (1791); Dresser, B. Europe, vii. p. 143, pl. 476 (1878); Hume and Marshall, Game B. of India, ii. p. 133, pl. (1879); et auctorum, *passim*.
Coturnix dactylisonans, Temm. Pig. et Gall. iii. pp. 478, 740 (1815); Gould, B. Europe, iv. pl. 263 (1837).
Coturnix vulgaris, Bout. Orn. Dauphiné, p. 72, pl. 43, fig. 1 (1843).
Coturnix coturnix, Licht.; Ogilvie-Grant, Cat. B. Brit. Mus. xxii. p. 233 (1893).

Adult Male.—General colour above sandy-brown, with pale buff shaft-stripes and black bars and markings; chin and throat *white, with a black anchor-shaped mark down the middle:* chest rufous-buff, with pale shafts; rest of under-parts paler. Total length, 6·7 inches ; wing, 4·2 ; tail, 1·5 ; tarsus, 1.

* The feathers under the wing, where it joins the body.

Adult Female.—Differs from the *male* in having no black band *down the middle* of the throat, and the chest more or less thickly spotted with brownish-black. From the *female* of *C. japonica* it may be readily distinguished by having the feathers on the chin and sides of the throat *short and rounded*.

The male described above is a typical example of *C. coturnix*. As considerable variation is to be found in the coloration of the chin and throat and their black markings, it may be as well to give here the substance of the remarks I have already published on this subject. The Migratory Quail has been constantly confused with two more or less resident local forms, *C. capensis*, found in South Africa, &c., and *C. japonica*, from Japan and China. The former is probably nothing more than a more richly coloured, rather smaller, resident local race of *C. coturnix*, but the latter is a perfectly distinct and easily characterised species. The migratory bird, wandering over an immensely wide range, visits the countries inhabited by both these forms, and constantly inter-breeds with them, the result being that all sorts of intermediate forms occur. The male of *C. japonica* has the chin and throat dull brick-red, devoid of any black markings, and the intermediate plumages between this species and the migratory bird are most noticeable among the *male* hybrids. For instance, some have the dull brick-red throat of *C. japonica* and the black anchor-shaped mark of *C. coturnix*; others have only the upper two-thirds of the throat dull red, and the lower third white; while, again, a third lot have, in addition, a black band down the middle of the red part; and all kinds of intermediate stages between these three examples may be found. These hybrids are, so far as I know, generally only found in Mongolia, China, and Japan, though there is one skin among the large series in the National Collection said to have been obtained in Bootan, N. India.

The Migratory Quail also inter-breeds freely with the chestnut-throated form (*C. capensis*) found in S. Africa and the islands surrounding the coast, and the results are to be seen in the many male birds from S. Africa and Southern Europe, &c., in which the white parts on the sides of the head and throat are more or less suffused with the bright rufous-chestnut characteristic of the resident bird.

A curious variety or semi-melanistic form of *C. coturnix* occurs in Spain in the marshy neighbourhood of Valencia. A male in the British Museum has the general colour of the plumage black, and the *female* has the under-parts suffused with sooty-brown.

Range.—Africa, Europe, and Asia, except in the south-east portion.

Habits.—The migratory habits of this species are well-known to most people, but though the great majority—countless hosts of Quail, which may be numbered by millions rather than thousands—shift their quarters in September and October, on the approach of winter, and move southwards, in many places a certain number remain and spend the winter where they have bred. For instance, in the South of England and Ireland, and in the countries bordering the Mediterranean, a few remain to winter, but the bulk of the European summer visitors betake themselves by various lines of migration to South Africa, from whence they return in March and April of the following spring. Enormous numbers also winter in India, crossing the Himalayas from Central Asia, while many arrive in Sind and Guzerat from the west, moving southwards from Beluchistan, Persia, and other northern latitudes.

The number of migrants varies greatly in different years, their movements being largely, if not entirely, regulated by the food-supply and seasonal conditions of the countries which they visit.

One may form some idea of the vast number to be met with in some parts of India from the following remarks by Tickell. He says: "In such localities as have been above noticed, Quails at times abound to such a degree that shooting them is mere slaughter. Where birds get up at every step dogs or beaters are worse than useless, and where the game is so plentiful, search after a wounded bird is seldom thought worth the trouble. It is usual to be provided with two or three guns,[*] to be loaded, as fast as emptied, by a servant. With one gun only it would be necessary to wash out the barrels two or three times in the course of the afternoon, or at all events to wait every now and then for them to cool. A tolerably good shot will bag fifty to sixty brace in about three hours, and knock down many others that are not found. I remember one day getting into a deyra, or island formed by alluvial deposit, in the Ganges, between Patna (Bankipore) and Sonepore, which was sown almost entirely over with gram (chunna), and which literally swarmed with Quail. I do not exaggerate when I say they were like locusts in number. Every step that brushed the covert sent off a number of them, so that I had to stand every now and then like a statue and employ my arms only, and that in a stealthy manner, for the purpose of loading and firing. A furtive scratch of the head, or a wipe of the heated brow, dismissed a whole "bevy" into the next field; and, in fact, the *embarras de richesse* was nearly as bad as if there had been no birds at all."

Nest.—A slight hollow in the ground, with little or no lining, and sheltered by standing crops or grass, &c.

Eggs.—Eight to twelve in number, sometimes more are laid; creamy-white or buff, more or less boldly blotched and spotted with rich brown. Average measurements, 1·15 by 0·88.

SUB-SP. *a*. THE CAPE QUAIL. COTURNIX CAPENSIS.

Coturnix capensis, Licht. *fide* Gray, Handl. B. ii. p. 263 (1870); Ogilvie-Grant, Ann. Mag. N. H. (6), ix. pp. 167, 169, 170 (1892); id. Cat. B. Brit. Mus. xxii. pp. 235, 237 (1893).

Adult Male.—Differs from the male of typical *C. coturnix* in having the sides of the head, chin, and throat bright rufous-

[*] He refers to the days before breech-loaders came in.

chestnut, and the mantle and chest washed with the same colour. It is also somewhat smaller. Total length, 6·3 inches; wing, 3·9-4; tail, 1·4; tarsus, 1·1.

Adult Female.—Very similar to the female of *C. coturnix*, but slightly smaller.

Range.—South Africa, south of about 15° S. latitude, Mauritius,[*] Madagascar, Comoro Islands, Cape Verd Islands, Canaries, Madeira, and the Azores.

II. JAPANESE QUAIL. COTURNIX JAPONICA.

Coturnix vulgaris japonica, Temm. and Schl. Faun. Jap. p. 103, pl. 61 (1842).

Coturnix japonica, Cass. in Perry's Exp. Jap. ii. p. 227 (1856); Prjevalsky, in Rowley's Orn. Misc. ii. p. 424 (1877); Ogilvie-Grant, Ann. Mag. N. H. (6), x. pp. 167, 170, 171 (1892; with woodcut of head of female); id. Cat. B. Brit. Mus. xxii. pp. 235, 239 (1893).

(*Plate XVI.*)

Adult Male.—Differs from the *male* of *C. coturnix* chiefly in having the sides of the head, chin, and throat *uniform dull brick-red*, with *no trace* of the black anchor-shaped mark, and the margins of the flank-feathers mostly rufous and much less spotted with black. Total length, 5·7 inches; wing, 3·9; tail, 1·2; tarsus, 1·05.

Adult Female.—Differs from the *female* of *C. coturnix* in having the feathers of the chin and throat *elongated and pointed*, especially on the sides, and generally margined with rufous; the chest and sides less spotted with black.

Young Males have the elongated throat-feathers as in the *adult female*, and the middle of the throat is suffused with dull brick-red. As the short, rounded, brick-red feathers of the adult are moulted, the elongated feathers disappear.

[*] There is a fine adult male of the typical Cape form in the National Collection said to have come from the Mauritius, but the locality *may* be a mistake. It is said that no indigenous Quail occurs there.

JAPANESE QUAIL.

Range.—Japan, South-east Mongolia, and China as far south as Canton. Specimens have also been obtained in Bootan, Northern India, and Karen-nee.

Habits.—Apparently very similar to those of the Migratory Quail.

B. *Outer web of the primary flight-feathers uniform brown.*

III. THE BLACK-BREASTED OR RAIN QUAIL. COTURNIX COROMANDELICA.

Tetrao coromandelicus, Gmel. S. N. i. pt. ii. p. 764 (1788).
Perdix coromandelica, Lath. Ind. Orn. ii. p. 654 (1790).
Coturnix coromandelica, Bonn., Tabl. Encycl. Méth. i. p. 221 (1791); Gould, B. Asia, vii. pl. 9 (1854); Hume and Marshall, Game Birds of India, ii. p. 152, pl. (1879); Oates, ed. Hume's Nests and Eggs Ind. B. iii. p. 444 (1890); Ogilvie-Grant, Cat. B. Brit. Mus. xxii. p. 241 (1893).
Coturnix textilis, Temm. Pig. et Gall. iii. pp. 512, 742 (1815).

Adult Male.—Like the *male* of *C. coturnix*, but the black pattern on the throat and neck is more strongly marked, and there is a large black patch covering the middle of the chest and breast. Total length, 5 inches; wing, 3·5; tail, 1·2; tarsus, 1·05.

Adult Female.—Very like the *female* of *C. coturnix*, but easily recognised by the absence of buff markings on the outer webs of the flight-feathers, as well as by its smaller size.

Range.—Greater part of the Peninsula of India, and extending to Assam, Manipur, Chittagong, and Pegu; also no doubt to Arakan.

Habits.—Generally speaking, the Rain Quail is merely a seasonal visitor over the greater part of its range, spending the monsoon in the drier parts of Upper and Western India, and the remainder of the year in the damp low-lying districts; but in many parts of Central India it is resident, merely shifting its feeding-ground with the change of season.

As a straggler it may sometimes be met with in the hills at an elevation of quite 6,000 feet, but the plains are its real home. Between the months of April and October, Mr. Hume says that it is habitually found in pairs, and singly during the cold season, while just after the young are able to fly, it may be found in coveys. The habits of this species are generally very similar to those of the Grey Quail, but the call is quite distinct, being a louder *double* (*not* a tri-syllabic) whistle.

According to Mr. Hume, "Rain Quail afford just as pretty shooting as the Common Quail when they are numerous; indeed, as they run less and fly rather faster, they yield perhaps better sport; but I have never known it possible to make such huge bags of these as one can of the other. In Upper India, during the winter and spring, you are pretty sure to pick up a brace or two along with the Grey Quail (with which they seem to associate on friendly terms) when shooting this latter; but I never knew more than five brace killed at this season in a day by one gun. But just when they first appear in the Doab in June or July, according as the rains are early or late, you may manage, by hard work, to get from twenty to thirty brace in a day, if you have steady dogs and there is plenty of grass about from two to three feet in height, or if, as is the case in some districts, there are a good many fields of the dwarf early rain millets."

Nest.—A slight hollow without lining, or with only a few blades of grass.

Eggs.—Average number about nine, sometimes more, often less. The ground-colour varies from yellowish-white to brownish-buff. The markings vary greatly in different clutches. Some are finely spotted and dotted all over with blackish or brown; others are heavily blotched and marked with rich brown, and much resemble those of the Migratory Quail; but numerous intermediate, and less heavily marked, sets are not uncommon. Average measurements, 1·09 by 0·83 inch.

IV. DELEGORGUE'S QUAIL. COTURNIX DELEGORGUEI.

Coturnix delegorguei, Deleg. Voy. Afr. Austr. ii. p. 615 (1847); Ogilvie-Grant, Cat. B. Brit. Mus. xxii. p. 243 (1893).

Coturnix histrionica, Hartl. Rev. et Mag. Zool. i. p. 495 (1849); id. Beitr. Orn. W.-Afr. pp. 1, 38, pl. xi. (1852).

Coturnix fornasini, Bianc. Spec. Zool. Mosamb. fasc. xvi. p. 399, pl. i. fig. 2 (1850).

Coturnix crucigera, Heugl. Vog. N. O.-Afr. p. 51 (1856).

Adult Male.—Easily distinguished from the male of *C. coturnix* by having the general colour of the under-parts chestnut, with a large black patch in the middle of the breast. Total length, 6·0 inches; wing, 3·7; tail, 1·3; tarsus, 1.

Adult Female.—Distinguished from the *female* of *C. coturnix* in having the general colour of the under-parts rufous-buff or dull chestnut.

Range.—Africa, south of about 15° N. latitude; recently obtained at Aden.

Very little is known about this rare Quail, but its habits are probably very similar to those of the common species.

V. THE AUSTRALIAN QUAIL. COTURNIX PECTORALIS.

Coturnix pectoralis, Gould, P. Z. S. 1837, p. 8; id. Syn. B. Austr. text and pl. fig. 1 (1837-8); North, Nests and Eggs B. Austr. p. 289 (1889); Ogilvie-Grant, Cat. B. Brit. Mus. xxii. p. 244 (1893).

Adult Male.—Differs from the *male* of *C. coturnix* chiefly in having the sides of the head, chin, and throat *dull* brick-red (as in *C. japonica*), but the feathers of the under-parts are white with black shaft-stripes, and there is *a black patch in the middle of the chest*. Total length, 7 inches; wing, 4·1; tail, 1·5; tarsus, 0·9.

Adult Female.—Differs chiefly from the *female* of *C. coturnix* in having the feathers of the chest and breast *longitudinally barred with black* near the extremity, the bars being *interrupted in the middle* by a wide buff interspace.

Range.—Australia and Tasmania.

Habits.—This is a very common bird all over Eastern and South-eastern Australia, as well as in Tasmania.

Gould writes: "Open grassy plains, extensive grass flats, and the parts of the country under cultivation are situations favourable to the habit of the bird; in its economy and mode of life, in fact, it so closely resembles the Quail of Europe (*C. coturnix*) that a description of one is equally descriptive of the other. Its powers of flight are considerable, and when flushed, it wings its way with arrow-like swiftness to a distant part of the plain; it lies well to a pointer, and has from the first settlement of the colony always afforded considerable amusement to the sportsman. It is an excellent bird for the table, fully equalling in this respect its European representative. . . . The chief food of this species is grain, seeds, and insects; the grain, as a matter of course, being only procured in cultivated districts, and hence the name of Stubble Quail has been given to it by the colonists of Tasmania, from the great numbers that visit the fields after the harvest is over."

Nest.—The slight nest of dry grass is placed in grassy flats or under some tuft of herbage on the open plain.

Eggs.—Seven to fourteen in number, varying considerably in markings, even in the same nest; usually with the ground-colour yellowish-white, with markings varying from minute freckles of umber-brown to large marbled blotches of a darker tint. Average measurements, 1·2 by 0·94 inch.

VI. THE NEW ZEALAND QUAIL. COTURNIX NOVÆ-ZEALANDIÆ.

Coturnix novæ-zealandiæ, Quoy and Gaimard, Voy. Astrol. Zool. i. p. 242, pl. 24, fig. 1 (1830); Buller, Hist. B. N. Zeal. i. p. 225, pl. xxiii. (1888); Ogilvie-Grant, Cat. B. Brit. Mus. xxii. p. 245 (1893).

Adult Male.—Like the male of *C. pectoralis*, but larger; the general colouring of the upper-parts warmer in tone; the sides of the head, chin, and throat *brighter* brick-colour or *chestnut*, the latter with a black bar on each side; and the fore-part of the neck is mostly black like the middle of the breast. Total length, 7·6 inches; wing, 4·6; tail, 1·8; tarsus, 1.

Adult Female.—Distinguished from the female of *C. pectoralis* by having the black bars on the chest- and breast-feathers *confluent* or very nearly so, and the black markings on the rest of the under-parts more numerous.

Range.—New Zealand.

This handsome species, once common in New Zealand, was long believed to be nearly, if not quite, extinct; for in 1888 Sir W. Buller remarked that no specimen had been heard of for the last twelve years. He says: "In the early days of the colony it was excessively abundant in all the open country, and especially on the grass-covered downs of the South Island. The first settlers, who carried with them from the old country their traditional love of sport, enjoyed some excellent Quail-shooting for several years; and it is a matter of local history that Sir D. Munro and Major Richmond, in 1848, shot as many as forty-three brace in the course of a single day within a few miles of what is now the city of Nelson, while a Canterbury writer has recorded that 'in the early days, on the plains near Selwyn, a bag of twenty brace of Quail was not looked upon as extraordinary sport for a day's shooting.'"

"It may be interesting to mention, as showing the value attaching to extinct or rapidly-expiring forms, that a skin of this bird (and that, too, a female) sent from the Canterbury Museum to Italy fetched as much as £75." The disappearance of this bird is no doubt largely due to the bush fires employed in clearing the Sheep-runs, as well as to the introduction of Dogs, Cats, and Rats.

It has since been ascertained that a few bevies still exist on the Kermadec Islands, but no doubt these will soon be exterminated for the sake of their market value.

Eggs.—Very similar to those of the Australian Quail; yellowish-brown **or** buff, thickly marked with spots and blotches **of** umber. One pair of eggs in the National Collection are so thickly spotted, that little of the ground-colour is visible, **while** another pair are not nearly so heavily marked. Measurements, about 1·3 by 0·9 inch.

THE SWAMP-QUAILS. GENUS SYNŒCUS.

Synoicus, Gould, B. Austr. v. pl. 89, or pt. xii. (1843).

Type. *S. australis* (Temm.).

Characters as in *Coturnix*, **but the axillary feathers are short and grey.**

Only three small species **are known.**

I. THE AUSTRALIAN SWAMP-QUAIL. SYNŒCUS AUSTRALIS.

Coturnix australis, **Temm.** Pig. et Gall. iii. pp. 474, 740 (1815).

Synoicus australis, Gould, B. Austr. v. pl. 89 (1843); North, Nests and Eggs Austr. B. p. 289 (1889).

Synoicus sordidus and *S. diemenensis*, Gould, P. Z. S. 1847, p. 33.

Synoicus cervinus, Gould, Handb. B. Austr. ii. p. 195 (1865).

Synœcus australis, **Ogilvie-Grant**, Cat. B. Brit. Mus. xxii. p. 247 (1893).

Adult Male.—The feathers of the upper-parts are reddish-brown, with **dull grey centres, the black mottlings** are few and fine, and the white **shafts, so conspicuous in** younger **birds, are scarcely visible.** The sides of the head and throat are dull grey. The feathers **of the under-parts are buff, with** grey centres **and almost** devoid of **black** cross-bars. *In somewhat younger examples* the plumage **of the** upper-parts is mottled **with** black, **and** barred **with** rufous, **the narrow** *white shafts* of the feathers being well defined; sides of the head and throat pale vinaceous-white; **rest of** under-parts buff, with V-shaped black cross-bars. Total length, about 7·5 inches; wing, 3·5–4·2;

tail, 1·7–2·0 ; tarsus, 0·9. This stage of plumage represents typical *S. australis* (Temm.). In *very old males* (*S. sordidus*, Gould) most of the markings on the upper- and under-parts disappear, and there is a general tendency to uniformity of colour in the plumage.

Adult Female.—Differs from the male in having the sides of the crown *black* or mostly black ; the black markings on the upper- and under-parts *much coarser ;* the centres of the feathers are *not grey ;* and the shaft-stripes are *buff* and *much wider* than in the *male*. From the female of *S. raalteni*, it is distinguished by having the chest pale rufous-buff, barred all over with black.

The Australian Swamp-Quail, as will be seen from the above list of names, was divided by Gould into four distinct species : but, from the large series I have examined, it is quite clear that the characters on which he relied are merely differences due to age and sex, and that all the forms are merely stages of plumage of one and the same species. It must also be noted that individuals vary, one from another, considerably in size, even in birds from the same locality, as may be seen from the measurements given above, though this is of very little importance. The changes in plumage of the upper-parts of the present species are very similar to those found in the Painted Quail (*Excalfactoria chinensis*).

Range.—South-eastern New Guinea, Australia, and Tasmania.

Habits.—This species is distributed all over Australia and Tasmania, and seems to prefer thick grassy flats and damp spots overgrown with undergrowth in the vicinity of rivers and water-holes. Gould says : " Its call is very similar to that of the Common Partridge, and, like that bird, it is found in coveys of from ten to eighteen in number, which simultaneously rise from the ground, and pitch again within a hundred yards of the spot whence they rose. It sits so close that it will often admit of being nearly trodden upon before it will rise. Pointers stand readily to it, and it offers, perhaps, better sport to the sportsman than any other bird inhabiting Australia. Its weight is about four ounces and three-quarters, and its flesh is delicious."

Nest.—Made of dry grass, &c., and placed on the ground among rank grass bordering running water.

Eggs.—From ten to eighteen in number; pale bluish-white, finely dotted all over with light brown. Average measurements, 1·17 by 0·92 inch.

II. RAALTEN'S SWAMP-QUAIL. SYNŒCUS RAALTENI.

Perdix raaltenii, Müll. and Schl. Land- en Volkenk. p. 158 (1839-44).

Coturnix raalteni, Wallace, P. Z. S. 1863, p. 486.

Synœcus raalteni, Ogilvie-Grant, Cat. B. Brit. Mus. xxii. p. 249 (1893).

Adult Male.—Differs chiefly from the most adult male of *S. australis* in having the sides of the head, chin, throat, and rest of under-parts *rufous*, with traces of dark cross-bars on the sides and flanks. As in less mature examples of *S. australis*, the *younger male* has the white shafts of the feathers of the upper-parts well marked, and the black bars on the under-parts stronger and extending over the breast and belly. Total length, 7·4 inches; wing, 3·5-3·8; tail, 1·5-1·7; tarsus, 0·95.

Adult Female.—Distinguished from the *male* by having the upper-parts blotched and marked with black, and the shaft-stripes wider and more distinct, while the under-parts are more strongly barred with black. From the *female* of *S. australis* it differs in having the chest pale dull rufous, with the black bars nearly obsolete.

Range.—Islands of Timor and Flores.

III. THE GREY SWAMP-QUAIL. SYNŒCUS PLUMBEUS.

Synœcus plumbeus, Salvadori, Ann. Mus. Civ. Genov. (2), xiv. p. 152 (1894).

Through the kindness of Count Salvadori, I have had the pleasure of examining the type of this species, which appears to be a very old male, and in general appearance resembles the most adult males of *S. australis*, but differs in having the plumage much greyer than that of any example of the latter species that I have seen.

Range.—South-east New Guinea.

THE PAINTED QUAILS. GENUS EXCALFACTORIA.

Excalfactoria, Bonap. C. R. xlii. p. 881 (1856).

Type, *E. chinensis* (Linn.).

Tail composed of only *eight* very short soft feathers, entirely hidden by the upper tail-coverts and less than half the length of the wing. The first primary flight-feather slightly shorter than, or sub-equal to, the second and longest.

Tarsi without spurs. Sexes entirely different.

All the three species of this genus are birds of extremely small size, and the plumage of the males is very beautiful.

I. THE COMMON PAINTED QUAIL. EXCALFACTORIA CHINENSIS.

The Chinese Quail, Edwards, Glean. Nat. Hist. v. p. 77, pl. 247 (1758).

Tetrao chinensis, Linn. S. N. i. p. 277 (1766).

Coturnix excalfactoria, Temm. Pig. et Gall. iii. pp. 516, 742 (1815).

Coturnix flavipes, Blyth, J. As. Soc. Beng. xi. p. 808 (1842).

Excalfactoria chinensis, Bonap.; Hume and Marshall, Game Birds of India, ii. p. 162, pl. (1879); Oates, ed. Hume's Nests and Eggs Ind. B. iii. p. 448 (1890); Ogilvie-Grant, Cat. B. Brit. Mus. xxii. p. 250 (1893).

Excalfactoria minima, Gould, P. Z. S. 1859, p. 128; id. B. Asia, vii. pl. vii. (1867).

Coturnix caineana, Swinh. Ibis, 1865, pp. 351, 542.

Adult Male.—Upper-parts brown, mottled and blotched with black, most of the feathers with whitish shaft-stripes, widest on the lower back and rump; forehead, sides of the head, and wing-coverts washed with dark slaty-blue, the latter mixed with bright chestnut; chin and throat handsomely marked with black and white; upper part of the chest, sides, and flanks slaty-blue; rest of the under-parts rich chestnut. Total length, 5·2 inches; wing, 2·8; tail, 1·1; tarsus, 0·8.

In *very old birds* the shaft-stripes on the upper-parts *entirely* disappear, and the whole aspect becomes darker and more uniform; on the under-parts the chestnut gradually takes the place of the slaty-blue colour till very little of the latter remains.

Younger Males.—The upper-parts are warmer brown, the black markings stronger, the shaft-stripes wider, and the under-parts are mostly slaty-blue, with only a small patch of chestnut on the middle of the belly.

Adult Female.—Upper-parts like those of the *younger male*, but the forehead and sides of the head are rufous-buff; the chin and throat white; and the under-parts buff, barred with black on the chest, sides, and flanks.

Range.—Ceylon, Indian Peninsula, and the Indo-Chinese countries; also Formosa, Celebes, Ternate. ? Hainan.

This extremely beautiful little Quail has a very wide distribution.

The somewhat darker and more strongly marked sub-species *E. lineata*, described below, is merely a southern representative of this bird, found in many of the larger islands of the Malay Archipelago and Australia, and it is extremely curious that we should find the typical *E. chinensis* in Celebes and Ternate. The examples collected by Mr. A. R. Wallace in the former island were described as a distinct species (*E. minima*) by Gould, and were supposed to differ from *E. chinensis* in being smaller, but even this distinction, slight as it is, does not hold good, for many examples from India and the Malay Peninsula are quite as small.

Habits.—Mr. Hume remarks: "I have always, except in the autumn, met with this species singly or in pairs. You may at times find a considerable number in the same patch of grass, but they are always as independent of each other as are similar aggregations of the Common Quail, and I totally disbelieve Latham's story of their going about in Sumatra in 'flocks of a hundred birds,' or in any sort of flocks or coveys except just after the breeding-season, when the two old birds, with their four to six young ones, *do* keep in a covey.

"Open, swampy, grassy lands or meadows are their favourite haunts, and I doubt whether they are ever found far from such. They will, doubtless, wander into low bush-jungle, the edges of low-standing crops, and, as Jerdon says, into patches of grass along the sides of roads; but this is almost exclusively when feeding in the early mornings and evenings, or when their meadow-homes have been suddenly flooded.

"They come freely into the open when feeding, and in the early mornings may be seen gliding along by the sides of roads and paths, picking about and scratching here and there, taking little notice of passengers, and either running on before them if not pressed, or just hiding up in the nearest tuft of grass, to emerge again as soon as the traveller has got ten or fifteen yards beyond their hiding-place.

"Their call is a very low, soft, double-whistled note, comparatively rarely heard except when a pair has been separated. Then, indeed, almost the moment the male has lit he begins calling to his mate. They feed quite silently, and, if they have *seen* and are expecting you, rise quite silently also; but both sexes, if suddenly alarmed, and females when startled from their nests, rise with a low, shrill, rapidly-repeated chirp, '*tchi, tchi, tchi.*' Their flight is very fast, straight, and low, rarely more than a foot above the tops of the grass, and is continued for from fifty to seventy yards, affording an excellent shot. Indeed, they fly so fast that, in places where they are abundant, they must, I should think, afford excellent sport. Always, be it understood, if you have small dogs to flush them; for without dogs, though you may or may not be able to start them at once, you will certainly not succeed in putting them up a second time.

"They feed chiefly on grass-seeds; very little, so far as my experience goes, on either grain or insects, though they do undoubtedly eat both of these. But I have always found them in meadows, where there was but little cultivation in the neighbourhood, and, perhaps, when they occur where millet-fields are common, they may, as I have been told, feed equally on these small grains. . . .

"This species is clearly monogamous. The *hen* sits (not the male, as in the Bustard Quails), and the male is always to be found near at hand; and when the young are hatched both parents accompany the brood for at least two months after they are able to fly.

"I have had reason to suspect that they may breed twice a year, but the matter is still doubtful, as the different periods at which we have found their nests may be due to differences in the climate of the localities in which we met with them."

Nest.—A mere depression in the ground, in a clump of coarse grass, loosely lined with a few grass-stems.

Eggs.—Five or six in number; rather broad ovals and with some gloss; olive-brown, more or less speckled with minute reddish-brown or purplish-grey dots. Average measurements, 0·98 by 0·76 inch.

SUB-SP. *a*. THE ISLAND PAINTED QUAIL. EXCALFACTORIA LINEATA.

La Petite Caille de l'Isle de Luçon, Sonnerat, Voy. N. Guin. p. 54, pl. 24 (1776).

Oriolus lineatus, Scop. Del. Flor. et Faun. Insubr. ii. p. 87 (1786).

Tetrao manillensis, Gmel. S. N. i. pt. ii. p. 764 (1788).

Excalfactoria chinensis, Auctorum, *passim; nec* Linn.

Excalfactoria australis, Gould, Handb. B. Austr. ii. p. 197 (1865).

Excalfactoria lineata, Ogilvie-Grant, Cat. B. Brit. Mus. xxii. p. 253 (1893).

Adult Male.—Differs from the *male* of *E. chinensis* in having the general colour of the upper-parts much darker and more strongly blotched with black.

Adult Female.—Distinguished from the *female* of *E. chinensis* by having much more black on the upper-parts, while the under-parts are darker and much more strongly barred with black.

Range.—Philippines, Palawan, Sulu Islands, Borneo, Java, Sumatra, and Australia.

II. THE NEW BRITAIN PAINTED QUAIL. EXCALFACTORIA LEPIDA.

Excalfactoria lepida, Hartlaub, Ber. Ver. Hamb. vii. November (1879); Ogilvie-Grant, Cat. B. Brit. Mus. xxii. p. 254 (1893).

Adult Male.—Upper-parts darker than in typical *E. chinensis*, and similar to those of *E. lineata*, but easily distinguished from both these forms by having no trace of chestnut on the wing-coverts, while the under-parts are entirely slaty-blue, except the lower part of the belly and under tail-coverts, which are chestnut. Total length, 4·8 inches; wing, 2·7; tail, 0·9; tarsus, 0·7.

Adult Female.—We have never had the opportunity of examining the female of this species, but it is probably very similar to that of *E. lineata*.

Range.—New Britain, New Ireland, and the Duke of York Group to the East of New Guinea.

III. ADANSON'S PAINTED QUAIL. EXCALFACTORIA ADANSONI.

Coturnix adansonii, Verr. Rev. et Mag. de Zool. 1851, p. 515; Sharpe, ed. Layard's Birds S. Afr. p. 606 (1884).

Excalfactoria adansonii, Bonap.; Ogilvie-Grant, Cat. B. Brit. Mus. xxii. p. 255 (1893).

Coturnix emini, Reichenow, J. f. O. 1892, p. 18, pl. 1, fig. 3 [male].

Adult Male.—Differs from the male of *E. chinensis* chiefly in having the upper-parts blackish-brown washed with slate; the upper tail-coverts and wing-coverts chestnut, the latter with slate-grey shaft-stripes; the under-parts dark slate-grey, except the sides and flanks, which are bright chestnut. Total length, 5.2 inches; wing, 2.9–3; tail, 1.1; tarsus, 0.8.

Younger Males have the middle of the back blotched with black, but in the more adult examples these marks disappear.

Adult Female.—Very similar to the *female* of *E. lineata*, but the wing-coverts are more strongly barred with black. The females of this species appear to average rather larger than the males, the wing measuring 3.1–3.2; but we have not examined a very large series of birds.

Range.—Africa, south of about 5° north latitude.

Habits.—Adanson's Painted Quail has a very wide distribution in Africa, being found in suitable localities over the greater part of that vast continent.

It is rather a rare bird, and its habits appear to be very similar to those of its eastern ally, *E. chinensis*.

THE PHEASANTS, TURKEYS, AND GUINEA-FOWLS (PHASIANINÆ).

The first flight-feather is considerably shorter than the tenth,* the tail is shorter or longer (often much longer) than the wing, and the sides of the head are feathered or entirely naked. [If the first flight-feather is longer than the tenth, the tail is *always* considerably longer than the wing.]

The most typical form of Pheasant-wing is found in the Argus Pheasant (*Argusianus argus*), where the first flight-feather is the *shortest*, and the tenth the *longest*.

THE STONE PHEASANTS. GENUS PTILOPACHYS.

Ptilopachus, Swainson, Class. B. ii. p. 344 (1837).

Type, *P. fuscus* (Vieill.).

Tail composed of *fourteen* feathers, rather long and rounded and more than three-fifths of the length of the wing.

First flight-feather somewhat shorter than the tenth; fifth slightly the longest.

Feet without spurs in either sex.

A large naked space behind the eye. Sexes similar in plumage.

Only one African species is known.

1. THE AFRICAN STONE PHEASANT. PTILOPACHYS FUSCUS.

Perdix fusca, Vieill. Tabl. Encycl. Méth. i. p. 366 (1823); id. Gal. des Ois. ii. p. 40, pl. ccxii. (1825); Jard. and Selby, Illustr. Orn. (new series), pl. xvi. (1837).

Perdix ventralis, Valenc. Dict. Sci. Nat. xxxviii. p. 435 (1825).

---

* The genus *Phasianus*, including the typical Pheasants, forms an exception, the first flight-feather being about equal to the *eighth*; but the length of the tail, which is always greater than that of the wing, at once distinguishes it as one of the *Phasianinæ*, though, as already remarked on p. 78, the distinction between this group and the *Perdicinæ* is a purely artificial one.

Ptilopachus erythrorhynchus, Swains. B. of W. Afr. ii. p. 220 (1837).
Ptilopachys fuscus, Ogilvie-Grant, Cat. B. Brit. Mus. xxii. p. 256 (1893).

Adult Male and Female.—Upper-parts brown, finely mottled with whitish; mantle and chest mostly sienna, with a dark shaft-band down the middle of each feather; sides of the head and throat dark brown, edged with white; middle of the breast uniform buff; belly dark brown; sides and flanks chestnut, with irregular cross-bars of brown and white.

Male measures: Total length, 11 inches; wing, 5·2; tail, 3·6; tarsus, 1·2.

Female: Smaller; wing, 4·7.

Range.—Africa, extending from Senegambia and the Gold Coast to Kordofan, Abyssinia, and the Sük country.

Habits.—Very little has been recorded about the habits of this curious bird, which seems to be met with chiefly on the bare stony hillsides at a considerable elevation.

The only account I can find is that given by Heuglin, and the substance of his remarks is as follows:—The Stone Pheasant is a gregarious bird, living in flocks of from five to fifteen individuals. It is only met with in rocky ground in the neighbourhood of cliffs and precipices, always in the proximity of running water or wells, and seems to prefer the neighbourhood of scrub and coarse grass. Flocks of these birds are apt to escape notice on account of the protective colour of their plumage, which harmonises perfectly with their surroundings, and renders them almost invisible. In the breeding-season, however, and throughout the rainy season, their presence is generally betrayed by the far-reaching flute-like whistle of the male; and in the early morning, and towards evening, one often falls in with a covey on their way to or from the water. The way in which these birds get over the rough stony faces of the hills reminds one of the Chukar (*Caccabis chukar*), for they hop from point to point in just the same way, helping themselves along with their wings. If surprised, they instantly disappear among the crevices in the rocks, and are then very difficult to flush; if in scrub or grass, they always prefer, if possible, to escape by running; but when pressed by a dog, they rise with a whirring flight and make for thick cover, where they are in the habit of drying their plumage after heavy rain or dew. Heuglin often noticed a peculiar habit of these birds during the breeding-season, from July to September, when whole flocks are wont to repair to some particular playground, usually a small bare spot sheltered by the bushes. The hens are more numerous than the cocks, and the former withdraw from the scene of action into the neighbouring cover, while the males strut round the open space, challenging and answering the chorus from neighbouring parties. Their note may be syllablised as *dŭī-dīŭ, dŭī-dīŭ, dŭī-dīŭ, dŭī-dīŭ, dī*, which is repeated at longer or shorter intervals.

Simultaneously the males commence dancing and showing off, ruffling their neck-feathers, nodding their heads, flirting their tails like a fan, and trailing their wings along the ground, while they circle round the playground with hops and springs. Half-grown young, still partially in the down, were often met with in January.

The Stone Pheasants generally roost under the shelter of overhanging rocks. Their flesh is said to be white and sweet, and Heuglin reckoned them one of the best of African Game-Birds.

Mr. F. J. Jackson, Captain Shelley, and Mr. T. E. Buckley, all state that they found this species in *pairs*, not in flocks; but perhaps the birds they met with were breeding, which may account for the difference between their observations and those of Heuglin, given above. The preponderance of the females over the males, combined with the curious habit indulged in by the Stone Pheasants, of repairing to some particular spot, where the males display their charms and pay court to the females, seem white, suggest that these birds are polygamous, like the True Pheasants; and certainly their behaviour during the pairing-season resembles that of the Black Game and other Grouse.

Nest.—Placed on the ground at the foot of a rock and hidden by coarse grass and scrub.

Eggs.—Yellowish-white; like miniature eggs of the Golden Pheasant.

THE BAMBOO-PHEASANTS. GENUS BAMBUSICOLA.

Bambusicola, Gould, P. Z. S. 1862, p. 285.

Type, *B. thoracica* (Temm.).

Tail composed of *fourteen* feathers, rather long and wedge-shaped, more than three-fourths of the length of the wing.

The first flight-feather is much shorter than the tenth, and the fifth is generally the longest.

Plumage of sexes similar. Males (and sometimes females) have a pair of spurs.

1. FYTCH'S BAMBOO-PHEASANT. BAMBUSICOLA FYTCHII.

Bambusicola fytchii, Anderson, P. Z. S. 1871, p. 214, pl. xi.; id. Zool. Res. Yun-nan, Birds, p. 673, pl. liv. (1878); Hume and Marshall, Game Birds of India, ii. p. 97, pl. (1879); Ogilvie-Grant, Cat. B. Brit. Mus. xxii. p. 257 (1893).

Bambusicola hopkinsoni, Godwin-Austen, P. Z. S. 1874, p. 44.

Adult Male and Female.—General colour above brown; nape mostly chestnut; feathers of the upper-back dark chestnut in the middle, and more or less mottled with white; wing-coverts strongly marked with buff, dark chestnut, and black; quills *mostly chestnut; eyebrow-stripes*, sides of the head, and throat buff; a black band from behind the eye down the side of the neck; chest brown, marked with chestnut and white; rest of under-parts buff, with heart-shaped black spots on the sides and flanks

Male measures: Total length, 12·3 inches, day fig, 5·8; tail, 4·4; tarsus, 1·9.

Female: Rather smaller.

Range.—North-eastern Bengal: Garo, Khasia, and Naga Hills in Assam; also the hills of North Cachar, East Manipur, and Yun-nan, and extending to South-western Sze-chuen and the Shan States.

Habits.—This Bamboo-Pheasant is a shy bird, frequenting dense grass and rarely met with in the open, except at dawn. When first flushed, they fly rapidly, often perching on trees, but never rise a second time if they can avoid doing so; their note, most often heard in spring, is, as one might expect from their affinities, somewhat fowl-like, and very different from that of the Tree Partridges, which are met with in similar localities. It is, according to Mr. Damant, who had opportunities of observing this bird in various parts of Assam, nowhere very common, and only found in the heavy forest-jungles at heights of not less than 2,500 feet, and most often in pairs; they are difficult to shoot, as they will not rise till hard pressed. Mr. Oates recently obtained specimens of this bird from the Shan States, and also a single egg.

Eggs.—The only example we have seen is almost perfectly oval in shape, the small end being but slightly pointed; colour uniform pale rufous-buff. Measurements, 1·45 by 1·1 inch.

II. THE CHINESE BAMBOO-PHEASANT. BAMBUSICOLA THORACICA.

Perdix thoracica, Temm. Pig. et Gall. iii. pp. 335, 723 (1815).
Perdix sphenura, G. R. Gray, Zool. Misc. p. 2 (1844); id. Fasc. B. China, pl. viii. (1871).
Arboricola bambusæ, Swinhoe, Ibis, 1862, p. 259.
Bambusicola thoracica, Swinhoe, P. Z. S. 1863, p. 307; Ogilvie-Grant, Cat. B. Brit. Mus. xxii. p. 258 (1893).

Adult Male and Female.—General coloration much like that of the Common Partridge (*Perdix perdix*). Above mostly olive-brown, marked with chestnut on the back and scapulars, and with some white and buff markings. Quills mostly blackish-brown; *eyebrow-stripes grey;* sides of the head and throat rufous-chestnut; chest mostly grey; rest of under-parts buff, with dark transverse spots on the sides and flanks; tail mostly chestnut.

Male: Total length, 11.8 inches; wing, 5.4; tail, 3.8; tarsus, 1.7.

Female: Somewhat smaller; wing, 5.2 inches.

Range.—South China, extending from Fokien to Sze-chuen and South Shen-si.

III. THE FORMOSAN BAMBOO-PHEASANT. BAMBUSICOLA SONORIVOX.

Bambusicola sonorivox, Gould, P. Z. S. 1862, p. 285; id. B. Asia, vi. pl. 63 (1864); Ogilvie-Grant, Cat. B. Brit. Mus. xxii. p. 259 (1893).

Adult Male and Female.—General plumage like that of *B. thoracica*, but richer and darker, and distinguished by having only the chin and throat chestnut, the *sides of the face being dark grey*, like the eyebrow-stripes and sides of the neck.

Male: Total length, 9.6 inches; wing, 5.1; tail, 3.5; tarsus, 1.5.

Female: Rather smaller.

Range.—Island of Formosa.

Habits.—Swinhoe gives the following account of the Formosan Bamboo-Pheasant:—"This and the Foochow Bamboo-fowl (*B. thoracica*) are of very similar habits and notes. This species is found throughout all the hills of Formosa, generally scattered about the bush, never in coveys. It is very pugilistic, the males and females both singing the same loud cry, beginning with *killy-killy,* and ending rapidly with *ke-put-kwai,* which is so powerfully uttered that it may be heard at a great distance. They are not easily flushed, lying so close to the ground that you may walk over the spot whence the noise appears to come, and rarely put up the bird. Each pair selects its own beat, setting up frequently during the day the challenge note, and woe betide any other Partridge that encroaches on the forbidden ground! They both set on him at once, and buffet him without mercy till he takes to his heels. This pugnacious propensity often meets, as perhaps it deserves to do, with an evil fate. The Chinese fowler listens for the challenge, and sets on the disputed hill a trap with a decoy within. The decoy is trained, and sets up a reply. The lord and lady of the manor rush to the spot and run recklessly into the trap and are caught. The captures are taken to the market and sold as cage-birds, the Chinese having a great love for the horrible screeching cry that this bird is incessantly sending forth. In the night this species leaves the shelter of the grass and bush, and repairs to the branches of bamboos and other trees to roost. It is an excellent percher, being quite at home on a branch, in which respect it differs from the Chinese Francolin (*Francolinus chinensis*), which never perches."

Nest.—A depression in the ground under the shelter of a bush or tuft.

Eggs.—Numerous; seven to twelve or more in number; dark brownish cream-colour, much like those of the Common Partridge (*P. perdix*). Measurements, 1·38 by 1·0 inch.

THE SPUR-FOWL. GENUS GALLOPERDIX.

Galloperdix, Blyth, J. As. Soc. Beng. xiii. pt. 2, p. 936 (1844).

Type, *G. lunulata* (Valenc.).

Tail composed of *fourteen* feathers, fairly long and rounded, the outer feathers being shorter than the middle pair.

The first flight-feather is much shorter than the second, which is about equal to the tenth; the fifth and sixth are rather the longest.

A large naked space round the eye.

Plumage of the sexes different. The feet of the male armed with two, and sometimes with three pairs of spurs. In the female one pair of spurs is usually developed, sometimes, but more rarely, two, though not unfrequently one or other of the feet may have two spurs.

I. THE RED SPUR-FOWL. GALLOPERDIX SPADICEA.

La Perdrix rouge de Madagascar, Sonnerat, Voy. Ind. Orient. ii. p. 169 (1782).
Tetrao spadiceus, Gmel. S. N. i. pt. ii. p. 759 (1788).
Francolinus spadiceus, J. E. Gray, Ill. Ind. Orn. ii. pl. 42, fig. 2 (1834).
Polyplectron northiæ, J. E. Gray, Ill. Ind. Orn. ii. pl. 43, fig. 1 (1834).
Ithaginis madagascariensis, G. R. Gray, List Brit. Mus. Gall. p. 32 (1844).
Galloperdix spadiceus, Blyth; Gould, B. Asia, vi. pl. 68 (1854); Hume and Marshall, Game Birds of India, i. p. 247, pl. (1878); Oates, ed. Hume's Nests and Eggs Ind. B. iii. p. 423 (1890); Ogilvie-Grant, Cat. B. Brit. Mus. xxii. p. 261 (1893).

Adult Male.—General colour brownish-chestnut or rufous-chestnut, most of the feathers with pale greyish-brown margins; *crown of the head dark brown;* sides of the head and neck greyish-brown. Total length, 14·6 inches; wing, 6·3; tail, 5·4; tarsus, 1·7.

Adult Female.—Differs in having the upper-parts irregularly barred with black and buff, and the feathers of the neck and under-parts tipped with black.

Specimens from Mt. Abu and the dryer northern parts of this bird's range, are paler and less strongly marked than examples from Southern India.

Range.—Peninsula of India, more especially the western parts. Madagascar [introduced].

As Mr. Hume very ably puts it :—" Certainly the distribution of the Red Spur-Fowl is as yet very imperfectly understood, and it inosculates so strangely with that of the Painted Spur-Fowl (*G. lunulata*), as will be seen when I come to deal with that species, that at present I can make nothing of the question. Both species seem to me to affect almost the same localities, and to have exactly the same habits, to be in fact complemental species, like the Red and Grey Jungle Fowl, or the Black and Painted Partridges, &c., and the way in which they seem to overlap each other's areas of distribution by many hundreds of miles is therefore most inexplicable. I need scarcely add that this species is essentially Indian, and occurs nowhere out of India.

Habits.—"The Red Spur-Fowl ranges from nearly sea-level to an elevation at Abu, the Pulneys, and the Nilgiris of 4,000 to 5,000 feet; indeed, on the latter it *has* been shot at over 7,500 feet. It is essentially a bird of forests and jungle, on hilly and broken land. It is unsafe to generalise from one's own limited personal experience, but I have the impression that the Red Spur-Fowl goes in more for forests and earth, and that the Painted one more affects scrub-jungle and *rocks*. You rarely, if ever, find the Red, you constantly find the Painted, Spur-Fowl in very rocky ground." (*A. O. Hume.*)

The late Mr. Davison, who was familiar with the species in the Nilgiris, says: "It seems to affect by preference dense and thorny cover in the vicinity of cultivation, but is also found in small isolated patches of jungle or sholas, and along the outskirts of the larger forests. It is perhaps found more numerously on the lower portions of the northern and western slopes of the Nilgiris.

"Though," as Dr. Jerdon remarks, "two or three Spur-Fowl usually form part of a day's bag on the Nilgiris, they are by no means easy birds to obtain; for without dogs it is almost impossible to flush them, and I have often observed that, even with dogs, they will run before these, till they come to some dense thorny bush, when they will silently fly up out of reach, and hide themselves in the thickest part, and once so concealed, it is almost impossible to flush them without cutting the bush to pieces. When flushed they rise with a cackle, and fly well and strong for a couple of hundred yards. Their flight is very like that of the 'Kyah Partridge.' They are usually found in small coveys of four or five birds, and when flushed do not rise together, but at irregular intervals, dispersing in different directions; they are often found in pairs, and not unfrequently I have come across single birds.

"They come into the open in the mornings and evenings to feed, and wander about a good deal. Even after they have retired into the shade they do not rest quietly, but wander hither and thither under the trees, scratching about among the dead leaves.

"A well-wooded ravine with plenty of thorny undergrowth, and with a stream of water in it, is always a favourite resort of this species.

"I do not think that this species is in any degree migratory, but no doubt, in many localities, in hot weather, when all springs and pools dry up, the birds shift their quarters a few miles to where water is available. With this exception, wherever it occurs, it is, I believe, a permanent resident, and there breeds."

There can be little doubt that this species is monogamous, as they are always found in pairs during the breeding-season.

Nest.—A slight hollow scratched in the ground and lined sparingly with dry leaves and grass, under the shelter of more or less dense undergrowth, generally in bamboo-thickets.

Eggs.—Four to seven in number, sometimes as many as ten; fowl-like; varying in colour from brownish- or pinkish-buff to cream-colour, and devoid of markings. Average measurements, 1·67 by 1·28 inch.

II. THE PAINTED SPUR-FOWL. GALLOPERDIX LUNULATA.

Perdix lunulata, Valenc. Dict. Sci. Nat. xxxviii. p. 446 (1825).

Perdix hardwickii, J. E. Gray, Ill. Ind. Zool. i. pl. 52 (1830-32).

Francolinus nivosus, Delessert, Mag. de Zool. Ois. pl. 18 (1840).

Galloperdix lunulosa, Gould, B. Asia, vi. pl. 69 (1854); Sclater, in Wolf's Zool. Sketches (2), pl. 41 (1861).

Galloperdix lunulatus, Hume and Marshall, Game Birds of India, i. p. 255, pl. (1878); Oates, ed. Hume's Nests and Eggs Ind. B. iii. p. 425 (1890); Ogilvie-Grant, Cat. B. Brit. Mus. xxii. p. 263 (1893).

Adult Male.—*Crown of the head black*, with some purplish-green gloss and *spotted with white;* upper-parts chestnut, with white black-edged spots; rest of head, throat, and neck white, spotted and barred with black; under-parts buff, spotted with black. Total length, 13·6 inches; wing, 6·2; tail, 4·8; tarsus, 1·5.

Adult Female.—Crown black, with chestnut shaft-stripes; above dull olive-brown, most of the feathers with dusky margins; eyebrow-stripes, sides of the head, and throat mostly chestnut; under-parts dull brownish-ochre shading into olive-brown; most of the feathers with a blackish marginal spot or band.* Total length, 12·6 inches; wing, 5·9; tail, 4·4; tarsus, 1·4.

Range.—Peninsula of India, especially the eastern portions.

Habits.—Although the Red Spur-Fowl and the present species inhabit much the same area, on the whole the latter may be said to be more of an eastern form, though the ranges of the two birds constantly overlap, and in many localities both species are met with.

Colonel Tickell writes: "In all places its skulking habits cause it to be very seldom seen. It haunts rocky places buried in thorny thickets, sometimes the stony jungly beds of nalas or small rivers, but more generally the isolated granite hills covered with dense brushwood, which are so common a feature in Chota Nagpore. It is generally in beating those

---

* In some examples these black spots are absent.

huge rocks with large bodies of men, when bear-shooting, that the 'Askal' is seen, and I have sometimes observed two or three in the air at a time, flying straight, with rapid action of the wings, much like Jungle Fowl. They are flushed but once; and after alighting, run into fissures and holes amongst the rocks, whence there is no dislodging them."

Captain Baldwin, again, says: "The male does not crow like the Jungle Cock, though both sexes make a kind of clucking noise like a true fowl. When running, these birds carry the tail up, not like a Partridge. I have often watched them when hidden behind a bush or rock, waiting for the beat to approach; sometimes over a dozen have run past me. They move very fast, and seldom take wing till hard-pressed. The flight is swift and rarely at any great height from the ground. The birds take a good hard blow to bring them down."

Nest.—None; the eggs being deposited on the bare ground sheltered by a rock or root of a tree, and concealed by surrounding tufts of grass.

Eggs.—Generally longer ovals than those of the Red Spur-Fowl, and uniform pale brownish-buff. Average measurements, 1·62 by 1·11 inch.

III. THE CEYLON SPUR-FOWL. GALLOPERDIX BICALCARATA.

Perdix bicalcaratus, Pennant, Ind. Zool. p. 40, pl. vii. (1769).
Perdix zeylonensis, Gmel. S. N. i. pt. ii. p. 759 (1788).
Galloperdix bicalcarata, Layard, Ann. Mag. N. H. (2), xiv.
    p. 105 (1854); Legge, Birds of Ceylon, iii. p. 741, pl.
    (1880); Hume and Marshall, Game Birds of India, i.
    p. 261, pl. (1878); Oates, ed. Hume's Nests and Eggs
    Ind. B. iii. p. 426 (1890); Ogilvie-Grant, Cat. B. Brit.
    Mus. xxii. p. 264 (1893).

Adult Male.—Crown of the head, neck, mantle, and sides black, with a wide white shaft-stripe to each feather; rest of upper-parts chestnut, with rather large white black-edged spots on the wing-coverts; chin and throat white; chest black, with a large white patch on each feather; rest of under-parts mostly white, edged with black, varying in width, according to age. Total length, 12 inches; wing, 6·1; tail, 4·5; tarsus, 2.

Adult Female.—Crown *blackish;* feathers of the forehead and sides of the head with pale rufous centres; chin and throat white; rest of the plumage *chestnut,* finely mottled with black. Total length, 10·8 inches; wing, 5·6; tail, 3·8; tarsus, 1·9.

Range.—Ceylon.

Habits.—To Colonel Legge's excellent work on the "Birds of Ceylon," I am indebted for the following note:—"The shy habits of this bird would prevent its being detected in most places where it is even abundant, were it not for its noisy cries or cackling, so well known to all who have wandered in our Ceylon jungles.

"It frequents tangled breaks, thickets in damp nalas, forest near rivers, jungle over hillsides, and in fact any kind of cover which will afford it entire concealment.

"It runs with great speed, and has a knack of noiselessly beating a retreat at one time, while at another it ventriloquises its exciting notes until the sportsman becomes fairly exasperated, and gives up the attempt he has made to stalk it in disgust. I have more than once endeavoured to cut off its retreat, or flush it by rushing into a little piece of jungle or detached copse in which I had found it, and from which it seemed impossible for it to escape, but I invariably failed in the attempt —a failure aggravated by my utter bewilderment at its unaccountable disappearance.

"The cock birds begin to call about six in the morning, and when one has fairly commenced, the curious ascending scale of notes is taken up from one to another until the wood resounds with their cries.

"They seem always to keep in small parties, which perhaps consist of the young of the year with their parents.

"The natives in the Central Provinces snare them with horse-hair nooses set in spots which they are observed to frequent in the early morning.

"They do not live well in confinement, either killing themselves by fighting or knocking their brains out by flying up against the top of their aviaries, and if they escape this fate, they are liable to die of some disease."

Nest.—None; situation similar to that chosen by the Painted Spur-Fowl.

Eggs.—Uniform cream-colour. Measurements, 1·42 to 1·43 by 1·12 inch.

THE PHEASANT-QUAIL. GENUS OPHRYSIA.

Ophrysia, Bonap. C. R. xliii. p. 414 (1856).

Type; *O. superciliosa* (Gray).

Tail composed of *ten* feathers, rather long and wedge-shaped, the outer pair being about two-thirds of the length of the middle pair.

First flight-feather much shorter than the tenth; fifth or sixth longest.

Plumage long and soft, and quite different in the two sexes.

The feet not provided with spurs in either sex.

Only one species of this genus is known, a small bird about the size of a Common Quail, but differing entirely from that species and all its group in most of its structural characters. I have no doubt that the nearest allies of this pigmy Pheasant—for that is really what it is—are the Blood Pheasants (*Ithagenes*) which follow. The rather stout coral-red bill, dull red feet, the long, soft, rather loose plumage, the shape of the wing, and the rather long tail are all characteristic of the Blood Pheasants, but not of the Quails. Unfortunately the present species is so rare, and so few examples have ever been obtained, that its anatomy has never been examined, but the probability is, that its bones would teach very little, for the skeletons of all the Quails, Partridges, and Pheasants are remarkably alike.

1. THE MOUNTAIN PHEASANT-QUAIL. OPHRYSIA SUPERCILIOSA.

Rollulus superciliosus, Gray, Knowls. Menag. Aves, p. 8, pl. xvi. (1846).

Ophrysia superciliosa, Bonap. C. R. xliii. p. 414 (1856); Hume and Marshall, Game Birds of India, ii. p. 105, pl. (1879); Ogilvie-Grant, Cat. B. Brit. Mus. xxii. p. 266 (1893).

Malacoturnix superciliosus, Blyth, P. Z. S. 1867, p. 475; Gould, B. Asia, vii. pl. 8 (1868).

Malacortyx superciliaris, Blyth, Ibis, 1867, p. 313.

Adult Male.—Middle of crown and nape brownish-grey, with black shaft-stripes; sides of crown black; forehead and a wide band down each side of the crown white; sides of the head, chin, and throat black, with a white band on each side of the latter; rest of upper- and under-parts grey, the former washed with olive-brown, and all the feathers edged with black; under tail-coverts black, tipped and spotted with white. Total length, 9 inches; wing, 3·5; tail, 3; tarsus, 1.

Adult Female.—Upper-parts brown, most of the feathers with black shaft-stripes or blotches; a black band on each side of the crown; eyebrow-stripes and sides of the head vinous-grey; throat whitish; under-parts similar to the back, but paler and more tawny. Total length, 8·8 inches; wing, 3·5; tail, 2·7; tarsus, 1.

Range.—North-western India, in the neighbourhood of Masuri and Naini Tal.

Habits.—This is still one of the least known of all the Indian Game-Birds, the total number of specimens recorded amounting to less than a dozen; and, so far as I am aware, no additional specimens have been obtained since the one shot by Major Carwithen near Naini Tal in 1876.

There can be little doubt that these birds are merely winter migrants from Tibet, though some occasionally remain till the beginning of summer. They rarely leave the cover of thick grass-jungle and brushwood, and cannot be flushed without the aid of dogs. When on the wing, their flight is slow and heavy, and, after going a short distance, they drop again into cover.

Those met with were generally in coveys of from six to ten, and found at elevations varying from 5,000 to 7,000 feet. When feeding on the fallen grass seeds, they utter a soft Quail-like note, but when separated, after they have been flushed, their call-note is a shrill whistle.

Captain Hutton says: "During the forenoon they wander up to feed amongst the long grass, to which they obstinately cling, feeding on the fallen seeds, and their presence being made known by their short Quail-like note. They will not come out into the open ground, and in the afternoon they descend into sheltered hollows amongst the grass and brushwood."

It is no doubt owing to the singularly retiring habits of this bird that so few specimens have as yet been obtained. We have several times tried to induce friends shooting in the neighbourhood of Masuri to look for and collect specimens, but so far without result, probably, as Mackinnon remarked, because these birds are very small, and involve an immense deal of bother in shooting, and when bagged, prove poor eating!

Nest and Eggs.—Nothing is known.

THE BLOOD PHEASANTS. GENUS ITHAGENES.

Ithaginis, Wagler, Isis, 1832, p. 1228.

Type, *I. cruentus* (Hardwicke).

Tail composed of fourteen feathers, rather long, about four-fifths of the length of the wing and slightly rounded, the outer feathers being somewhat shorter than the middle pair.

First flight-feather much shorter than the second, which is about equal to the tenth; fifth rather the longest.

Bill very short and stout. A large naked patch round the eye.

Plumage long and soft. Male with a full crest, and the feathers of the body pointed.

PLATE XVII

BLOOD-PHEASANT

Feet of the male armed with two or more pairs of spurs; females devoid of these appendages or with a pair of blunt knobs.

Plumage quite different in the two sexes.

Only three species are known.

I. THE BLOOD PHEASANT. ITHAGENES CRUENTUS.

Phasianus cruentus, Hardwicke, Tr. Linn. Soc. xiii. p. 237 (1822) [male].

Phasianus gardneri, Hardwicke, Tr. Linn. Soc. xv. p. 167 (1827) [female].

Ithaginis cruentus, Wagler, Isis, 1832, p. 1228; Elliot, Monogr. Phasian. ii. pl. 30 (1872); Hume and Marshall, Game Birds of India, p. 155, pl. (1878).

Ithagenes cruentus, Ogilvie-Grant, Cat. B. Brit. Mus. xxii. p. 268 (1893).

(*Plate XVII.*)

Adult Male.—Crown buff or rufous-buff; upper-parts grey, with white shaft-stripes, washed with green on the wings; longer median wing-coverts *green;* upper tail-coverts widely margined with crimson; forehead and feathers round the eye black; chin, throat, and cheeks *crimson;* rest of the under-parts shading into pale *green*, darkest on the sides and belly, the feathers of the chest and breast being more or less edged with crimson.* Under tail-coverts crimson, tipped with greenish-white. Total length, 15·6 inches; wing, 8·3; tail, 6·8; tarsus, 2·6.

Adult Female.—*Forehead*, chin, and throat† *rust-colour;* back of the head and nape slate-grey; upper-parts pale brown, under-parts reddish-brown, all finely mottled with darker colour. Total length, 11·5; wing, 7·7; tail, 5·7; tarsus, 2·3.

Range.—Higher regions of Nepal, Native Sikhim, Sikhim, and Western Bootan; it also extends into Tibet.

* The crimson edges are most marked in birds from Nepal, much less so, or absent, in examples from Sikhim.

† I have seen an example in which the chin and throat are washed with crimson; perhaps a barren female beginning to assume male plumage.

The peculiar grass-green colour characteristic of the males of this genus is not seen in any other species of Game-Bird. The only other bird of this Order with green plumage is the female of the Red-Crested Wood Partridge (*Rollulus roulroul*), but in this instance the colour is much darker.

Habits.—Mr. Hume publishes the following notes by Hodgson, which give some idea of the bird's habits:—"This species is common in Nepal in flocks of twenty to thirty in the same situations as the Moonal, that is to say, in the higher forests and in the immediate neighbourhood of the snow, even outside, though always near, the forests.

"They greatly affect the clumps of Mountain Bamboo, and feed about on the ground amongst these, much like domestic fowls, turning over the leaves and grasses with their feet, scratching about in the ground, and picking up insects, grass, seeds, grain, and wild fruit.

"They do not eat the bulbous roots of which the Moonal is so fond. On any alarm the whole flock utter a sharp alarm-note (*ship, ship*), and scuttle away.

"In the winter the birds come southward a little, but never approach the Great Valley. Numbers are caught in November and December, and in their own haunts they are by no means rare. Packs are often seen consisting of as many as seventy to one hundred birds. They ascend and descend with the snow, and are easily captured, being fearless and stupid. They prefer somewhat inaccessible places. Their flight is short and feeble."

Sir J. Hooker, who met with the Blood Pheasant in Eastern Nepal and Sikhim at elevations of from 10,000 to 14,000 feet, remarks: "During winter it appears to burrow under or in holes amongst the snow; for I have snared it in January in regions thickly covered with snow, at an altitude of 12,000 feet. I have seen the young in May. The principle food of the bird consists of the tops of the pine and juniper in spring, and the berries of the latter in autumn and winter; its flesh has always a very strong flavour, and is moreover uncommonly tough; it was, however, the only bird I obtained at those great elevations in tolerable abundance for food, and that not very frequently. The Bhutias say that it acquires an additional spur every year; certain it is that they are more numerous than in any other bird, and that they are not alike on both legs. I could not discover the cause of this difference; neither could I learn if they were produced at different times. I believe that five on one leg, and four on the other, is the greatest number I have observed."

Mr. W. T. Blanford adds, in his notes on the zoology of Sikhim:—"All that I saw were in the pine-forests round Yeomatong, where they were tolerably abundant. They rarely take flight even when fired at, but run away and often take refuge on branches of trees. I have shot five or six out of one flock by following them up; they usually escape uphill, and if, as frequently takes place, the flock has been scattered, after a few minutes they commence calling with a peculiar long cry, something like the squeal of a Kite. The only other note I heard was a short monosyllabic note of alarm; I have heard a bird utter this when sitting on a branch within twenty yards of me.

"In their crops I found small fruits, leaves, seeds, and in one instance what appeared to me to be the spore-cases of a moss; there were no leaves or berries of juniper, and the birds were excellent eating. We did not notice the unpleasant flavour mentioned by Hooker, probably because better food is abundant at the season when we shot our birds, and they consequently do not then feed upon pine or juniper."

Nest and Eggs.—Nothing definite is known of the Blood Pheasant's nesting habits, but the nest, loosely constructed of grass and leaves, is said to be placed on the ground among grass and bushes, and to contain ten to twelve eggs.

II. GEOFFROY'S BLOOD PHEASANT. ITHAGENES GEOFFROYI.

Ithaginis geoffroyi, Verr. Bull. Soc. d'Acclim. (2), iv. p. 706 (1867); Gould, B. Asia, vii. pl. 42 (1872); Elliot, Monogr. Phas. ii. pl. 31 (1872); David and Oustalet, Ois. Chine, p. 401, pl. 113 (1877).

Ithagenes geoffroyi, Ogilvie-Grant, Cat. B. Brit. Mus. xxii. p. 269 (1893).

Adult Male.—Differs chiefly from the male of *I. cruentus* in having the long crest-feathers *grey*, with white shafts; a larger patch of green on the wing-coverts; and the *chin*, *throat*, and chest *grey*. Total length, 17 inches; wing, 7·7; tail, 6; tarsus, 2·6.

Adult Female.—Differs from the female of *I. cruentus* in having the *forehead*, sides of the head, chin, and throat *brownish*; the upper- and under-parts *alike* greyish-brown, the latter finely mottled all over with blackish-brown. Total length, 16 inches; wing, 7·1; tail, 5·1; tarsus, 2·5.

Range.—Higher regions of Eastern Tibet and Western Szechuen, China.

Habits.—Writing of this species, Abbé David, its original discoverer, says that it lives in more or less numerous flocks near the limits of the upper forest-region, preferring the bamboo-jungles. Ordinarily its food consists of young shoots, leaves, and seeds, but the stomachs of three birds he killed in April, whilst the country was still covered with snow, contained absolutely nothing but moss. These fine birds are in the habit of perching on trees, and they are extremely sociable by nature, and after the young are hatched, several old pairs in company bring up their united families and form one covey.

Nest.—One found on the ground under brushwood in the forest, at an elevation of 13,500 feet above sea-level.

Eggs.—Buff, spotted with reddish-brown. Average measurements, 1·95 by 1·31 inch.

III. THE NORTHERN BLOOD PHEASANT. ITHAGENES SINENSIS.

Ithaginis sinensis, David, Ann. Sci. Nat. (5), xviii. art. 5, p. 1 (1873), and xix. art. 9, p. 1 (1874); id. and Oustalet, Ois. Chine, p. 402, pl. 114 (1877).

Ithaginis geoffroyi, Prjev. (*nec* Verr.), Mongolia, ii. p. 122, (1876); id. in Rowley's Orn. Misc. ii. p. 421 (1877).

Ithagenes sinensis, Ogilvie-Grant, Cat. B. Brit. Mus. xxii. p. 270 (1893).

Adult Male.—Differs from *I. geoffroyi* in having the sides of the crest *blackish-brown*, and the patch on the wing-coverts *rust-brown*; chin, throat, and fore-neck blackish-grey with whitish shaft-stripes, washed on the chin with crimson. Total length, 17·6 inches; wing, 8; tail, 7; tarsus, 2·5.

Adult Female.—Differs from the female of *I. geoffroyi* in having the upper-parts browner, more like those of *I. cruentus*, but paler; the throat is *dirty grey* and the breast pale brownish-buff, with *scarcely a trace* of dark mottlings. Total length, 16·2 inches; wing, 7·5; tail, 5·9; tarsus, 2·4.

Range.—Higher regions north of the Nan-shan and Kan-su Mountains, also the Sinling Mountains between Shen-si and Ho-nan.

Habits.—Prjevalsky says: "We observed this scarce species, called by the natives 'Sermun,' only in the Kan-su Mountains, where it principally inhabits the wooded districts, and also ascends to the alpine regions. We did not obtain a single specimen ourselves, but bought a skin from the Tanguts, who told us that these birds, in spring, keep mostly to the edges of forests and about the alpine bushes, and then feed on a particular kind of grass. In winter they descend to the middle and low mountain ranges, where they form small companies, and pass the night on trees like *Crossoptilon auritum*.

"The note of the present species consists of a long, perfectly clear, but not loud whistle."

THE HORNED PHEASANTS. GENUS TRAGOPAN.

Tragopan, Cuvier, Règ. Anim. éd. 2, i. p. 479 (1829).

Type, *T. satyra* (Linn.).

Tail composed of *eighteen* feathers, rather long and wedge-shaped, the outer pair being about two-thirds of the length of the middle pair.

First flight-feather shorter than the tenth and much shorter than the second; the fourth or fifth rather the longest.

Axillary feathers very long.

Sides of the head nearly naked or thinly feathered in the males, completely so in the females. The male has a short crest, an elongate, fleshy, erectile horn inserted above each eye, and a large gular flap or apron-like wattle, most prominent in the breeding-season, and especially when the birds are excited by passion, but scarcely visible in winter. Feet armed in the male (rarely in the female) with a pair of short, stout spurs.

Plumage of sexes quite different.

1. THE CRIMSON HORNED PHEASANT. TRAGOPAN SATYRA.

Horned Indian Pheasant, Edwards, Nat. Hist. B. iii. pl. 116 (1750).

Meleagris satyra, Linn. S. N. i. p. 269 (1766).

Phasianus cornutus, P. L. S. Müll. Natursyst. Suppl. p. 125 (1776).

Tragopan satyrus, Cuv. Règne Anim. i. p. 479 (1829); Gould, Cent. B. Himal. pl. 62 (1832); Temm. Pl. Col. v. pls. 13, 14 [Nos. 543, 544] (1834); Ogilvie-Grant, Cat. B. Brit. Mus. xxii. p. 271 (1893).

Satyra pennanti, pl. 49, and *S. lathami*, pl. 51, J. E. Gray, Ill. Ind. Zool. i. (1830-32), and *S. nepaulensis*, id. *t.c.* ii. pl. 40 (1834).

Ceriornis satyra, G. R. Gray; Gould, B. Asia, vii. pl. 49 (1868); Elliot, Monogr. Phas. i. pl. 22 (1872); Hume and Marshall, Game Birds of India, i. p. 137, pl. (1878); Oates, ed. Hume's Nests and Eggs Ind. B. iii. p. 409 (1890).

coloured or salmon on the sides, which are spotted or edged with blue. Total length, 27 inches; wing, 10·8; tail, 9·2; tarsus, 3·1.

Adult Female.—Differs from the female of *T. satyra* in having the general tone of the plumage much greyer, with very little rufous-buff, even on the wings and under-parts.

Range.—Higher ranges of the western Himalayas from Native Gurhwal westwards to Cashmere.

Habits.—Writing from Kulu, of this species, which is commonly, though incorrectly, known as the "Argus" by most Indian sportsmen, Mr. Young remarks: "They keep in companies of from two or three to ten or a dozen, not in compact flocks, but scattered widely over a considerable space of forest, so that many at times get quite separated, and are found alone. . . .

"The trees furnishing them with a sufficiency of food, though the ground be covered with snow many feet in depth, the severest storms of winter do not, speaking of the species generally, cause them to change their locality. After a severe fall of snow, a few occasionally leave for a time their usual haunts, if in a very bleak quarter, or at any considerable elevation, and are found in places widely differing, as small patches of forest on a bare exposed hillside, narrow wooded ravines, patches of low brushwood and jungle, and anywhere where the ground is sheltered from the sun by trees and bushes. Sometimes one is found in a similar situation in fine weather, probably driven out of its retreat by an Eagle[*] or Falcon; but these are rare exceptions, and they soon again return to their regular resorts.

"At this season, except for its note of alarm when disturbed, the *Tewar* is altogether mute, and is never heard of its own accord to utter a note or call of any kind, unlike the rest of our Pheasants, all of which occasionally crow or call at all seasons. When alarmed, it utters a succession of wailing cries, not un-

[*] The Nepal Hawk-Eagle (*Limnaëtus nipalensis*) is an inveterate foe to both species of Tragopan and to the Moonal.

like those of a young lamb or kid, like the syllables "waa, waa, waa," each syllable uttered slowly and distinctly at first, and more rapidly as the bird is hard pressed or about to take wing. . . .

"In spring, as the snow begins to melt on the higher parts of the hill, they leave entirely their winter resorts, and gradually separate and spread themselves through the more remote and distant woods, up to the region of birch and white rhododendron, and almost up to the extreme limits of forest.

"Early in April they begin to pair; and the males are then more generally met with than at any other period; they seem to wander about a great deal, are almost always found alone, and often call at intervals all day long. When thus calling, the bird is generally perched on the thick branch of a tree, or the trunk of one which has fallen to the ground, or on a large stone. The call is similar to the one they utter when disturbed, but is much louder, and only one single note at a time, a loud energetic "waa," not unlike the bleating of a lost goat, and may be heard for upwards of a mile. It is uttered at various intervals, sometimes at every five or ten minutes for hours together, and sometimes not more than two or three times during the day, and most probably to invite the females to the spot.

"When the business of incubation is over, each brood, with the parent birds, keep collected together about one spot, and descend towards their winter resorts as the season advances; but the forests are so densely crowded with long weeds and grass, that they are seldom seen till about November, when the vegetation has partially decayed and admits of a view through the wood.

"They feed chiefly on the leaves of trees and shrubs: of the former, the box and oak are the principal ones; of the latter, *ringal* and a shrub something like privet. They also eat roots, flowers, grubs and insects, acorns and seeds, and berries of various kinds, but in a small proportion compared with leaves. In confinement they will eat almost any kind of grain.

"Though the most solitary of our Pheasants, and in their native forests perhaps the shyest, they are the most easily reconciled to confinement; even when caught old they soon loose their timidity, eating readily out of the hand; and little difficulty is experienced in rearing them."

Nest.—Placed on the ground, and roughly constructed of grass, small sticks, and a few feathers.

Eggs.—Six in number (in the one nest found); long ovals, pointed at the smaller end, with very little gloss but fine shell; pale buff, very finely granulated with a darker shade. Average measurements, 2·51 by 1·7 inches.

III. TEMMINCK'S HORNED PHEASANT. TRAGOPAN TEMMINCKI.

Satyra temminckii, J. E. Gray, Ill. Ind. Zool. i. pl. 50 (1830-32).
Ceriornis temmincki, G. R. Gray, Gen. B. iii. p. 499 (1845);
    Sclater, List of Phas. p. 11, pl. 11 (1863); Gould, B. Asia, vii. pl. 46 (1869); Elliot, Monogr. Phasian. i. pl. 24 (1872); David and Oustalet, Ois. Chine, p. 118, pl. 112 (1877).
Tragopan temmincki, Ogilvie-Grant, Cat. B. Brit. Mus. xxii. p. 275 (1893).

Adult Male.—Differs chiefly from the two last-mentioned species in having the occipital crest orange-red, the upper-parts Indian red, with *pearl-grey* spots edged with black, and the under-parts dark *Indian red*, with a large *grey spot* near the extremity of each feather. Horns blue; gular flap deep blue, barred with red on the outer margins. Total length, 25 inches; wing, 9·9; tail, 7·8; tarsus, 3·2.

Adult Female.—Apparently much like that of *T. satyra*, but we have never had the opportunity of examining specimens, except those living in the aviaries at the Zoological Gardens.

Range.—South-western and Central China, extending from the Mishmi Hills through Sze-chuen to Southern Shen-si and Hoo-pih.

Habits.—Abbé David tells us that this bird lives a solitary life on the wooded mountains, seldom leaving the thick covert, and feeding on seeds, fruits, and leaves. Its cry is very loud, and most nearly imitated by the syllable *oua* two or three times repeated, whence its Chinese name *Oua-oua-ky*, but it is also called *Ko-ky*, or *Kiao-ky*, meaning Horned Fowl, and *Sin-tsion-ky*, or Starred Fowl, on account of the grey spots adorning the plumage. The flesh is said to be capital eating. I am informed that this bird is not met with under about 10,000 feet above the sea-level.

IV. BLYTH'S HORNED PHEASANT. TRAGOPAN BLYTHI.

Ceriornis blythi, Jerd. P. As. Soc. Beng. 1870, p. 60; Sclater P. Z. S. 1870, pp. 163, 219, pl. 15; Gould, B. Asia, vii. pl. 47 (1872); Elliot, Monogr. Phasian. i. pl. 26 (1872); Hume and Marshall, Game Birds of India, i. p. 152, pl. (1878); Godwin-Austen, P. Z. S. 1879, p. 457, pl. xxxix.

Tragopan blythii, Ogilvie-Grant, Cat. B. Brit. Mus. xxii. p. 276 (1893).

Adult Male.—Head, neck, and chest orange-red; rest of upper-parts like those of *T. satyra*, but with a very dark red patch on each side of the white spot; sides and flanks similar; breast and belly *smoky-grey or greyish-buff*. Horns azure; orbital skin orange; gular flap brimstone, tinged with greenish-blue at the base. Total length, 24 inches; wing, 10·2; tail, 7·4; tarsus, 3·2.

Adult Female.—Like the female of *T. satyra*, but with the upper-parts blacker and less ferruginous; the lower-parts paler and without ferruginous-buff. From the female of *T. melanocephalus* it is distinguished by having the black and buff marking of the upper-parts much richer and darker. We have only seen living female examples of this species, and have had no opportunity of examining them close at hand.

Range.—Higher ranges of North-eastern Assam, east of the Burrail range and southwards to North-east Manipur.

Habits.—Mr. G. Damant writes of the "Grey-bellied Tragopan," as he calls it:—"This bird is found on most of the

PLATE XVIII.

CABOT'S HORNED PHEASANT

high ranges in the Nága Hills, notably on the Burrail range, near the villages of Kohima, Khenomah, and Mozemah.

"It is a permanent resident, and does not appear to migrate.

"It is found on the highest peaks (which attain an altitude of 9,000 feet in the Burrail range) and probably never descends to a lower elevation than 5,000 feet. It is said to breed in the month of April, and to lay three or four eggs.

"During the cold weather it is found at lower elevations than in the rains, as it descends as the mountain springs dry up.

"It appears to be generally distributed, but is not very common. Two live examples, now in my possession, eat worms and a kind of red berry very greedily. So far as I have observed, it has only one note resembling the syllable 'ak.'

"The Nágas catch these birds by laying a line of snares across a ravine which they are known to frequent, and then, with a large semicircle of beaters, driving the birds down to them. They go as quietly as possible so as not to frighten the birds sufficiently to make them take flight, as, if not much alarmed, they prefer running."

V. CABOT'S HORNED PHEASANT.   TRAGOPAN CABOTI.

Ceriornis caboti, Gould, P. Z. S. 1857, p. 161; id. B. Asia, vii. pl. 48 (1858); Elliot, Monogr. Phasian. i. pl. 25 (1872); David and Oustalet, Ois. Chine, p. 419, pl. 111 (1877).

Ceriornis modestus, David, MS.; David and Oustalet, Ois. Chine, p. 419 (1877).

Tragopan caboti, Ogilvie-Grant, Cat. B. Brit. Mus. xxii. p. 277 (1893).

(*Plate XVIII.*)

Adult Male.—Upper-parts, sides, and flanks differ from those of *T. blythi* in having each feather *black* down the middle, with a *buff* spot at the extremity and an Indian-red patch on each side; the basal part of each feather also spotted with white; under-parts *buff;* naked sides of head and gular flap reddish-orange, the latter marked on the sides and base with emerald-green. Total length, 23 inches; wing, 9·2; tail, 6·8; tarsus, 2·9.

Adult Female.—We have never been able to examine a female example except in aviaries, but it appears to resemble the female of *T. temmincki*.

Range.—South-eastern China; mountains between Fo-kien and Kiang-si. ? Also the hills in the interior of Quang-si.

Habits.—Abbé David found this somewhat aberrant species fairly common in the chain of mountains separating Fo-kien from Kiang-si. It is known to the natives by the same local names as *T. temmincki*, which bird it closely resembles in its habits, and its flesh is equally excellent for the table. Of the many specimens he examined in October and November not a single male was seen in female plumage, though at that season one would expect to find the young males of the year in that garb, and David came to the conclusion that this species differed from all the other members of the genus in getting its fully adult plumage at the first moult.

THE MOONAL PHEASANTS. GENUS LOPHOPHORUS.

Lophophorus, Temm. Pig. et Gall. ii. p. 355 (1813).

Type, *L. refulgens* (Temm.).

Tail composed of *eighteen* feathers, moderately long (shorter than the wing) and rounded, the outer pair being somewhat shorter than the middle pair.

The first flight-feather considerably shorter than the tenth, the fifth slightly the longest.

Male with an elongate crest of semi-upright spade-shaped plumes or with the top of the head (in *L. sclateri*) covered with curled feathers. A nearly naked space round the eye, and the feet armed with a stout spur.

Sexes quite different in plumage; most of the upper-parts in the male brilliantly metallic.

1. THE COMMON MOONAL PHEASANT. LOPHOPHORUS REFULGENS.

Lophophorus refulgens, Temm. Pig. et Gall. ii. p. 355 (1813), iii. p. 673 (1815); Ogilvie-Grant, Cat. B. Brit. Mus. xxii. p. 278 (1893).

Lophophorus impeyanus, Gould (*nec* Latham), Cent. B. Himal. pls. 60, 61 (1832); id. B. Asia, vii. pl. 53 (1850); Elliot, Monogr. Phasian. i. pl. 18 (1872); Hume and Marshall, Game Birds of India, i. p. 125, pl. (1878); Oates, ed. Hume's Nests and Eggs Ind. B. iii. p. 407 (1890).

Adult Male.—Top and sides of the head and crest composed of spade-shaped feathers, metallic-green shot with purplish-blue; back of the neck reddish copper-colour, shading into golden-green; mantle shining golden-green; wings mostly purplish-blue, changing to bronze-crimson; *lower back pure white;* under-parts black, with *no* green gloss except on the throat; tail light rufous-chestnut. Total length, 26 inches; wing, 11·6; tail, 9; tarsus, 3·2.

Adult Female.—Short crest, top of the head, mantle, rump, chest, and sides of breast black, with buff centres, mostly with black lines on each side of the shaft, and with irregular black bars and mottlings on the wings; lower back with more or less concentric irregular bars of black and buff; chin and throat white; rest of under-parts mottled with black and buff, and generally with distinct whitish shaft-stripes; tail black, barred with rufous. Total length, 23 inches; wing, 10·5; tail, 7·5; tarsus, 2·6.

Range.—Elevated forests of the Himalayas, from Eastern Afghanistan to Western Bhotan.

Every author writing since 1832 has followed Gould's original mistake in calling this bird *Lophophorus impeyanus*, a name which, without a shadow of doubt, Latham applied to the next species. He clearly states in his description that his bird had the *back* and wing-coverts *rich purple*, tipped with green-bronze, and the under-parts glossed with green. A glance at his figure shows that his *L. impeyanus* was not the present species, but the bird afterwards described as *L. chambanus* by Col. C. H. T. Marshall.

Habits.—Mr. Hume writes: "What is essential to this species is elevation and forest. All our Pheasants in the Himalayas may, as Hodgson (I think) pointed out thirty or forty years ago, be roughly divided into three classes: firstly, those of the high mountains, to which belong the Moonal, the Snow-Cocks, the Blood Pheasant, and the Tragopans; secondly, those of the mid-region, the Cheer, the Koklass, and the various Kalij Pheasants; and thirdly, the Jungle Fowl of the lower region.

"And you must have vegetation and forest as well as considerable altitudes; it would be vain to seek the Moonal in the stony wildernesses of Lahoul and Spiti, or the desert steppes of Ladakh. I have shot many Moonal in my time, and have seen a vast number more. There are few sights more striking, where birds are concerned, than that of a grand old cock shooting out horizontally from the hillside just below one, glittering and flashing in the golden sunlight, a gigantic rainbow-tinted gem, and then dropping stone-like, with closed wings, into the abyss below."

From the full and excellent account of this species given by Mr. Frederic Wilson I extract the following. He says:—

"The Moonal is found on almost every hill of any elevation, from the first great ridge above the plains to the limits of forest, and in the interior it is the most abundant of our Game-Birds. When the hills near Mussooree were first visited by Europeans, it was found to be common there, and a few may still be seen on the same ridge eastwards from Landour.

"In summer, when the rank vegetation which springs up in the forest renders it impossible to see many yards around, few are to be met with, except near the summits of the great ridges jutting from the snow, where morning and evening, when they come out to feed, they may be seen in the open glades of the forest and on the green slopes above. At that time no one would imagine they were half so numerous as they really are; but, as the cold season approaches, and the rank grass and herbage dies away, and they begin to collect together, the woods seem full of them, and in some places hundreds may be put up in a day's walk.

"In summer, the greater number of the males, and some of the females, ascend to near the limits of the forests where the hills attain to a great elevation, and may often be seen on the grassy slopes a considerable distance above these limits.

"In autumn, they all descend into the forest, frequenting those parts where the ground is thickly covered with decayed leaves, under which they search for grubs; and they descend lower and lower as winter sets in and the ground becomes frozen or covered with snow. . . .

"The females keep more together than the males; they also descend lower down the hills, and earlier and more generally leave the sheltered woods for exposed parts or the vicinity of the villages on the approach of winter. Both sexes are often found separately in considerable numbers. On the lower part or exposed side of the hill, scores of females and young birds may be met with, without a single old male; while higher up, or on the sheltered side, none but males may be found. In summer they are more separated, but do not keep in individual pairs, several being often found together.

"It may be questioned whether they do pair or not in places where they are at all numerous; if they do, it would appear that the union is dissolved as soon as the female begins to sit, for the male seems to pay no attention whatever to her whilst sitting, or to the young brood when hatched, and is seldom found with them.

"The call of the Moonal is a loud, plaintive whistle, which is often heard in the forest at daybreak or towards evening, and occasionally at all hours of the day.

"In severe weather numbers may be heard calling in different quarters of the wood before they retire to roost. The call has a rather melancholy sound, or it may be that, as the shades of a dreary winter's evening begin to close on the snow-covered hills around, the cold and cheerless aspect of Nature, with which it seems quite in unison, makes it appear so.

"From April to the commencement of the cold season, the Moonal, though there is nothing of cunning or artifice in its nature, is rather wild and shy, but this gives way to the all-taming influence of winter's frosts and snows; and from October it gradually becomes less and less wild, until it may be said to be almost tame, but as it is often found in places nearly free from underwood, and never attempts to escape observation by concealing itself in the grass or bushes, it is perhaps sooner alarmed, and at a greater distance, than other Pheasants, and may, therefore, appear to a casual observer at all times a little wild and timid. . . .

"It gets up with a loud fluttering, and a rapid succession of shrill screeching whistles, often continued till it alights, when it occasionally commences its ordinary loud and plaintive call and continues it for some time.

"In winter, when one or two birds have been flushed, all within hearing soon get alarmed; if they are collected together, they get up in rapid succession; if distantly scattered, bird after bird slowly gets up, the shrill call of each as it rises alarming others still farther off, till all in the immediate neighbourhood have risen. In the chestnut-forests, where they often collect in large flocks, and where there is little underwood, and the trees, thinly dispersed and entirely stripped of their leaves, allow of an extensive view through the wood, I have often stood till twenty or thirty have got up and alighted in the surrounding trees, and have then walked up to the different trees and fired at those I wished to procure without alarming the rest, only those very close to the one fired at being disturbed at each report. . . .

"The females appear at all times much tamer than the males. The latter have one peculiarity not common in birds of this Order: if intent on making a long flight, an old male, after flying a short way, will often cease flapping his wings, and soar along with a trembling vibratory motion at a considerable height in the air, when, particularly if the sun be shining on his brilliant plumage, he appears to great advantage, and certainly looks one of the most magnificent of the Pheasant-tribe.

"In autumn the Moonal feeds chiefly on a grub or maggot which it finds under the decayed leaves; at other times on roots, leaves, and young shoots of various shrubs and grasses, acorns, and other seeds and berries. In winter it often feeds in the wheat and barley fields, but does not touch the grain; roots and maggots seem to be its sole inducement for digging amongst it. At all times and in all seasons it is very assiduous in the operation of digging, and continues at it for hours together. In the higher forests, large open plots occur quite free from trees or underwood, and early in the morning, or towards evening, these may often be seen dotted over with Moonals, all busily engaged at their favourite occupation.

"The Moonal roosts in the larger forest-trees, but in summer, when near or above their limits, will often roost on the ground on some steep rocky spot. The flesh is considered by some nearly equal to Turkey, and by others as scarcely eatable. In autumn and winter many, particularly females and young birds, are excellent, and scarcely to be surpassed in flavour or delicacy by any of the tribe, while from the end of winter most are found to be the reverse."

Mr. Hume adds, "Once or twice late in April I have come upon males nautching, with wings drooped, tail cocked and outspread, and breast almost touching the ground, shivering and quivering spasmodically, and moving backwards and forwards with tiny steps like Turkey-cocks, but the birds were always off before I could really study the peculiarities of their nuptial dance."

Nest.—A hollow in the ground, sheltered by some rock, bush, or the root of a large tree; little or no lining.

Eggs.—Four or five, sometimes six, in number; oval and pointed towards the small end; pale whitish-buff, more or less thickly freckled all over, except towards the ends, with reddish-brown. Average measurements, 2·55 by 1·78 inches.

SUB-SP. *a.* LOPHOPHORUS MANTOUI.

Lophophorus impeyanus, var. *mantoui*, Oustalet, Bull. Soc. Zool. France, xviii. p. 19 (1893).

Adult Male.—Said to differ from the male of *L. refulgens* in having no trace of bronze-red on the neck, the interscapular region *purple*, and the black under-parts *slightly glossed with green*.

SUB-SP. *b.* LOPHOPHORUS OBSCURUS.

Lophophorus impeyanus, var. *obscura*, Oustalet, Bull. Soc. Zool. France, xviii. p. 19 (1893).

Adult Male.—Said to differ from the male of *L. refulgens* in having the head, crest, neck, and mantle very deep green, shading in some lights into black, and the wing-coverts, secondaries, and upper tail-coverts greenish-bronze, with some purple, black, and green reflections.

These two forms were recently founded by Dr. Oustalet on a couple of trade-skins obtained from a dealer who purchased them in the London market. Nothing is known respecting the locality they come from or any other particulars. Dr. Oustalet assures us that the colour of the feathers *cannot* have been chemically changed; if he is correct in this statement, these birds, especially the former, may represent some really distinct form of which we at present know nothing, but it is much more probable that both these examples are merely accidental varieties picked out from among the thousands of ordinary Moonal-skins that are annually imported into the London market.

II. THE IMPEYAN OR CHAMBA MOONAL PHEASANT. LOPHO-
PHORUS IMPEYANUS.

Impeyan Pheasant, Latham, Gen. Syn. Suppl. i. p. 208, pl. 114 (1787).
Phasianus impejanus (sic), Latham, Ind. Orn. ii. p. 632 (1790).
Phasianus curvirostris, Shaw, Mus. Lever. p. 101, pl. (1792).
Lophophorus impeyanus, v. Pelz. Ibis, 1873, p. 120; Ogilvie-Grant, Cat. B. Brit. Mus. xxii. p. 280 (1893).
Lophophorus chambanus, Marshall, Ibis, 1884, p. 421.

(*Plate XIX.*)

Adult Male.—Differs chiefly from *L. refulgens* in having the feathers of the *lower back golden-green, shading into purplish-blue towards their extremities;* upper tail-coverts chestnut, tipped with golden-green, and the under-parts entirely glossed with metallic golden-green. Total length, 26 inches; wing, 11·5; tail, 9; tarsus, 3.

Adult Female.—Unknown.

Range.—Chamba, N.W. Himalayas.

Remarks.—Although to Latham the credit of originally describing the male of this species undoubtedly belongs, Col. C. H. T. Marshall may at least claim the honour of having rediscovered this splendid bird, which had long been overlooked owing to the unanimity with which ornithologists united *L. impeyanus*, Latham, with *L. refulgens*, Temminck. Latham's type has unfortunately disappeared, and we have been unable to find any trace of it, though it at one time formed part of the collection in the Leverian Museum in London, and was the same individual described by Shaw as *Phasianus curvirostris*. Most of this collection, which was sold by auction in London in 1806, was purchased by the Vienna Museum, but Latham's type of *L. impeyanus* is no longer to be found.

Colonel C. H. T. Marshall, who re-discovered this species, which had been quite lost sight of since it was originally described by Latham in 1787, writing in the "Ibis" for 1884, remarks: "Two years ago a Monál Pheasant was brought in to me from the Birnota Forest (in the Chamba State, N.W. Himalaya), which I saw at once was very different from *L. impeyanus* (meaning *L. refulgens*). Its bronzed lower back and green breast made it easily distinguishable from any other known species. My brother, Colonel George Marshall, R.E., who was with me, suggested that I should describe it then, but fearing that it might be a mere variety, I considered it best to wait until more specimens could be procured. The following spring Mr. A. L. Seale (to whom I have given a contract to shoot Monál and Argus for skins in Chamba during the season) told me that he had had three specimens of what he called 'the Black-backed Monál' brought in to him from the same direction that my bird came from. On comparison I found that they agreed exactly with mine. This being, I consider, sufficient proof that it is a distinct species, I propose for it the name of *Lophophorus chambanus*, after the Raja of Chamba, in whose territories it was discovered."

The *female* has not yet been discovered, but it is greatly to be hoped that some of the many sportsmen, who go into Cashmere on shooting trips, will visit Chamba and secure examples of both sexes of this rare Moonal.

III. DE L'HUYS'S MOONAL PHEASANT. LOPHOPHORUS L'HUYSII.

Lophophorus l'huysii, Verr. and Geoffr. St.-Hil. Bull. Soc. Acclim. (2), iii. p. 223, pl. (1866), iv. p. 706 (1867); Sclater, P. Z. S. 1868, p. 1, pl. i.; Elliot, Monogr. Phasian. i. pl. 19 (1872); Gould, B. Asia, vii. pl. 54 (1873); David and Oustalet, Ois. Chine, p. 403, pl. 110 (1877); Ogilvie-Grant, Cat. B. Brit. Mus. p. 281 (1893).

Adult Male.—Like *L. refulgens*, but differs chiefly in having the crest purplish-bronze and composed of ordinary elongate feathers; lower back *white*; the rump-feathers *metallic golden-*

green, margined with white, and the tail bluish-green, glossed with purplish-blue; the middle of the feathers *mostly black*, irregularly *spotted* on each side of the shaft *with white*. Total length, 30 inches; wing, 12·6; tail, 10·3; tarsus, 3·3.

Adult Female.—Easily recognised from the *female* of *L. refulgens* by having the whole of the lower back pure white.

Range.—Western Sze-chuen in West China, extending to Eastern Koko-nor.

Habits.—Abbé David tells us that "this splendid Moonal inhabits the highest regions of Moupin and Eastern Koko-nor, as well as the western frontier of Sze-chuen, where it is met with in small flocks on the grassy slopes above the region of forest, roosting in the trees at night. Its general food consists of vegetable substances, particularly succulent roots, which it digs up with ease by the help of its strong beak. As it searches in particular for those of a *Fritillaria* commonly known as *Paè-mow*, the natives call it by the name of *Paè-mow-ky*. In its native country the adult male is also called *Ho-than-ky* (Shining Metallic-Fowl) on account of its metallic plumage. It is a very shy bird, of extremely powerful flight, and its cry, which one hears in the early morning and during rain, consists of three or four separate piercing notes uttered at intervals." From certain information that Abbé David received, he believed that this Moonal is also found in Yunan and in Quei-chow, and it is certain, in any case, that it is found throughout the greater part of Eastern Tibet, but it is everywhere rare, and it cannot be long before it completely disappears: for the Chinese are constantly in pursuit of it, and catch these splendid birds by means of snares for the sake of their delicate flesh.

This bird is found at a higher altitude than any of the other species, being met with on the rocky plateaux near the limit of perpetual snow, at elevations of about 16,000 feet above the sea-level. It roosts on the stunted rhododendrons or descends to the pine-forests.

IV. SCLATER'S MOONAL PHEASANT. LOPHOPHORUS SCLATERI.

Lophophorus sclateri, Jerdon, Ibis, 1870, p. 147, and J. As. Soc. Beng. 1870, p. 61; Sclater, P. Z. S. 1870, p. 162, pl. xiv.; Elliot, Monogr. Phasian. i. pl. 20 (1872); Hume and Marshall, Game Birds of India, i. p. 136, pl. (1878); Godwin-Austen, P. Z. S. 1879, p. 681, pl. li.; Ogilvie-Grant, Cat. B. Brit. Mus. xxii. p. 282 (1893).

Chalcophasis sclateri, Gould, B. Asia, vii. pl. 55 (1873).

Adult Male.—Top of the head covered with curly golden-green feathers, changing into blue; mantle and wings mostly steel-green, changing into purple; lower back, rump, and upper tail-coverts *white*, the two former with black shaft-stripes; tail *chestnut, with a wide white band at the extremity*, and the basal part of the feathers black, barred and mottled with buff. Total length, 26 inches; wing, 11·8; tail, 8·2; tarsus, 3·1.

Adult Female.—Chiefly distinguished from those already described by having the lower back *pale ochraceous-white, finely mottled with dark brown;* tail black, with six or seven narrow whitish-cross bars, and tipped with the same colour.

Range.—Hills to the east and south-east of Sadiya, in the extreme north-east of Assam.

Very few specimens have been obtained of this extremely scarce Moonal, and most, if not all, of the known examples have been brought down by the hill-tribes (Mishmis and Abors) to the fair held annually at Sadiya, the most easterly station in Assam.

THE CRESTLESS FIRE-BACKED PHEASANTS. GENUS ACOMUS.

Acomus, Reichenb. Nat. Syst. Vög. p. xxx. (1852).

Type, *A. erythrophthalmus* (Raffl.).

Tail composed of fourteen feathers, rather short and laterally compressed, or hen-like; the third pair being somewhat longer than the central ones, and very much longer than the outer pair.

First flight-feather considerably shorter than the second, which is about equal to the tenth; sixth rather the longest.

No crest in either sex.

A large naked red patch on each side of the head.

Feet, in both sexes, armed with a stout pair of spurs. Female black.*

I. THE MALAYAN CRESTLESS FIRE-BACK. ACOMUS ERYTHROPHTHALMUS.

Phasianus erythrophthalmus, Raffles, Tr. Linn. Soc. xiii. p. 321 (1822).

Phasianus purpureus, J. E. Gray, Ill. Ind. Zool. i. pl. 42 (1830-32) [female].

Euplocamus erythrophthalmus, Sclater, in Wolf's Zool. Sketches (2), pl. 34 (1861); Sclater, List of Phasian. p. 7, pl. 8 (1863); Elliot, Monogr. Phasian. ii. pl. 28 (1872).

Acomus erythrophthalmus, Ogilvie-Grant, Cat. B. Brit. Mus. xxii. p. 283 (1893).

Adult Male.—General colour of plumage *black*, glossed with purplish and steel-blue, and finely mottled with white; lower back fiery bronzy-gold, shading into bronzy-red on the rump; sides finely mottled with white; tail pale rufous-buff. Total length, 20 inches; wing, 9·5; tail, 6·4; tarsus, 3.

Adult Female.—Plumage entirely black, glossed with purplish or steel-blue. Total length, 18·5 inches; wing, 8·4; tail, 5·4; tarsus, 2·8.

Range.—Southern part of the Malay Peninsula and Sumatra. Has been recorded from Java, but probably in error.

Habits.—Practically nothing has been recorded about the habits of this bird, and the only examples obtained are those snared by natives. It is only known that the Malayan Crestless Fire-Back frequents the dense damp forests, and we may fairly assume that its habits are much like those of its ally the Crested Fire-Back, described below.

* The female of *A. inornatus* is still unknown.

The first examples of this species were obtained in Sumatra by Sir S. Raffles, and skins are generally to be found in collections of birds made by native collectors in the vicinity of Malacca, where it would seem to be fairly common.

Nothing is known of the eggs or nidification of this species.

II. THE BORNEAN CRESTLESS FIRE-BACK. ACOMUS PYRONOTUS.

Euplocomus erythrothalmus (sic), J. E. Gray (*nec* Raffles), Ill. Ind. Zool. ii. pl. 38, fig. 1 (1834).

Alectrophasis pyronota, G. R. Gray, List Gall. Brit. Mus. p. 26 (1844).

Euplocomus pyronotus, Elliot, Monogr. Phasian. ii. pl. 29 (1872).

Acomus pyronotus, Ogilvie-Grant, Cat. B. Brit. Mus. xxii. p. 284 (1893).

Adult Male.—Distinguished from the male of *A. erythrophthalmus* in having the neck and mantle grey, finely mottled with black, and with white shafts; the chest and breast black, *with white shaft-stripes*. Total length, 20 inches; wing, 9·3; tail, 5·7; tarsus, 3·3.

Adult Female.—Quite similar to the *female* of *A. erythrophthalmus*. Total length, 18·5 inches; wing, 8·4; tail, 5; tarsus, 3·15.

Range.—Sarawak, Borneo.

Mr. C. Hose tells us that this species is a low-country bird but is decidedly rare, and that its native name is "Singgier."

III. THE BLACK CRESTLESS FIRE-BACK. ACOMUS INORNATUS.

Acomus inornatus, Salvadori, Ann. Mus. Civ. Genov. xiv. p. 250 (1879); id. P. Z. S. 1879, p. 651, pl. xlviii.; Büttikofer, Notes Leyd. Mus. ix. p. 77 (1887); Ogilvie-Grant, Cat. B. Brit. Mus. xxii. p. 285 (1893).

Adult Male.—Much like the females of the last two species, being *entirely black*, but all the feathers of the upper-parts are distinctly and sharply edged with shining dark bluish-green, producing a scaled appearance. Total length, 18·5 inches; wing, 8·9; tail, 6·5; tarsus, 2·8.

Adult Female.—Has not yet been obtained.

Range.—Mount Singalan and the highlands of Padang. Western Sumatra.

Remarks.—I was at first inclined to believe that the male of this most interesting species, discovered by Dr. Beccari, had been wrongly sexed, in spite of that naturalist's assertions to the contrary. This was certainly not the case, for a second undoubted *male* example, perfectly similar to the type, has since been obtained by Dr. C. Klaesi, and is now in the Leyden Museum.

It is exceedingly remarkable that the male of this species should so closely resemble the females of the other species, and it will be extremely interesting to see, when the female of the Black Crestless Fire-Back (in this case somewhat of a misnomer) is discovered, whether it is black or reddish-brown, as Dr. Beccari imagined. He relied on the testimony of natives and on feathers which had been found near their traps, which belonged to specimens which had unfortunately been eaten by some carnivorous animal. It seems likely that these feathers may have belonged to females of *Lophura rufa*, in which, as we shall see below, the plumage and tail-feathers nearly answer to the description "reddish-brown."

The native name for this bird is said to be *Ajam merah mata*.

THE CRESTED FIRE-BACKED PHEASANTS. GENUS LOPHURA.

Lophura, Fleming, Philos. Zool. ii. p. 230 (1822).
Type, *L. rufa* (Raffles).

Tail rather long, composed of sixteen feathers, laterally compressed as in the Fowls; the third pair somewhat longer than the middle ones, and very much longer than the outer pair.

First flight-feather shorter than the second, which is about equal to the tenth; fifth and sixth slightly the longest.

A large naked red or blue patch on each side of the head, and a large wattle of the same colour on each side of the throat.

Male with a full crest, composed of more or less long bare shafts, with a bunch of plumes at the tip. Feet armed with a pair of stout spurs (absent in the female).

I. THE MALAYAN CRESTED FIRE-BACK. LOPHURA RUFA.

Phasianus ignitus, Raffles (*nec* Shaw), Trans. Linn. Soc. xiii. p. 320 (1822) [male]; Vieillot, Tabl. Encycl. Méth. i. p. 363, pl. 237, fig. 2 (1823).

Phasianus rufus, Raffles, Trans. Linn. Soc. xiii. p. 321 (1822) [female].

Euplocamus vieilloti, Gray, List Gen. B. 2nd ed. p. 77 (1841); Gould, B. Asia, vii. pl. 15 (1852); Sclater and Wolf, Zool. Sketches (2), pl. 36 (1867); Hume and Marshall, Game Birds of India, i. p. 213, pl. (1878).

Euplocamus ignitus, Elliot, Monogr. Phasian. ii. pl. 26 (1872).

Euplocamus sumatranus, Dubois, Bull. Ac. Belg. (2), xlvii. p. 825 (1879).

Lophura rufa, Ogilvie-Grant, Cat. B. Brit. Mus. xxii. p. 286 (1893).

Adult Male.—General plumage, including the crest, back, and *under-parts*, *black*, beautifully glossed with purplish-blue; lower back and rump fiery bronzy-red; feathers of the *sides and flanks with white* (or sometimes chestnut) *shaft-stripes**; middle pairs of tail-feathers *white;* naked sides of the head and wattles bright smalt-blue; feet bright red. Total length, 27 inches, wing, 11·6; tail, 10·2; tarsus, 4·3.

Adult Female.—General colour above chestnut, redder and darker *on the neck* and finely mottled with black; feathers of

---

* In some examples, especially in Sumatran birds (the *Euplocamus sumatranus*, Dubois, quoted above), the shaft-stripes are rufous-buff or chestnut instead of white, but this difference is not dependent on locality, and is apparently of no specific importance.

neck and chest chestnut, edged on the sides with white; those of the breast and sides of the belly black, usually mottled with chestnut and margined with white. Tail *dark chestnut;* naked sides of the head and feet like those of the male, but paler. Total length, 24 inches; wing, 10; tail, 7·6 ; tarsus, 3·6.

Range.—Siam and Southern Tenasserim southwards, the Malay Peninsula, and Sumatra.

Habits.—The late Mr. W. Davison, who is probably the only European who ever shot this bird in a wild state, says : " These birds frequent the thick evergreen forests in small parties of five or six ; usually there is only one male in the party, the rest being females, but on one or two occasions I have seen two males together ; sometimes the males are found quite alone. I have never heard the males crow, nor do I think that they ever do so ; when alarmed, both males and females have a peculiar sharp note, exceedingly like that of the large Black-backed Squirrel (*Sciurus bicolor*). The males also continually make a whirring sound with their wings, which can be very well imitated by twirling rapidly between the hands a small stick, in a cleft of which a piece of stiff cloth has been transversely placed. I have often discovered the whereabouts of a flock by hearing this noise. They never come into the open, but confine themselves to the forests, feeding on berries, tender leaves, and insects and grubs of all kinds, and they are very fond of scratching about after the manner of domestic poultry, and dusting themselves. When disturbed, they run rapidly away, not in different directions, but all keeping much together ; they rise at once before a dog, getting up with a great flutter, but when once well on the wing, fly with a strong and rapid flight; they seldom alight again under a couple of hundred yards, and usually on the ground, when they immediately start running.

" I noticed on one occasion a very curious thing. I had stalked an Argus, and while waiting to obtain a good shot.

I heard the peculiar note, a sort of '*chukun, chukun,*' followed by the whirring noise made by the male Fire-Back, and immediately after saw a fine male Fire-Back run into the open space, and begin to chase the Argus round and round its clearing. The Argus seemed loath to quit its own domain, and yet not willing to fight, but at last, being hard pressed, it ran into the jungle. The Fire-Back did not attempt to follow, but took up a position in the middle of the clearing, and recommenced the whirring noise with his wings, evidently as a challenge, whereupon the Argus slowly returned, but the moment it got within the cleared space the Fire-Back charged it, and drove it back into the jungle, and then, as before, took up his position in the middle of the space and repeated the challenge. The Argus immediately returned, but only to be again driven back, and this continued at least a dozen times, and how much longer it would have continued I cannot say, but a movement on my part attracting the birds' attention, they caught sight of me, and instantly, before I could fire, disappeared into the jungle. The Argus never made the slightest attempt to attack the Fire-Back, but retreated at once on the slightest movement of the latter towards it, nor did I see the Fire-Back strike the Argus with either bill, wings, or spurs."

Nest.—Nothing is known of the nidification.

Eggs.—An egg laid in confinement, in July, is pale brownish-buff, like that of the Game-Fowl, but larger, with little or no gloss, and covered with minute pores.

II. THE BORNEAN CRESTED FIRE-BACK. LOPHURA IGNITA.

Phasianus ignitus, Shaw, Nat. Misc. ix. pl. 321 (*c.* 1787).
Gallus macartneyi, Temm. Pig. et. Gall. ii. p. 273 (1813); iii. p. 663 (1815).
Euplocomus nobilis, Sclat. P. Z. S. 1863, p. 118, pl. xvi.; Elliot, Monogr. Phasian. ii. pl. 27 (1872).
Lophura ignita, Ogilvie-Grant, Cat. B. Brit. Mus. xxii. p. 288 (1893).

Adult Male.—Differs chiefly from the *male* of *L. rufa* in having the *lower breast and upper belly fiery bronzy-gold*, and the middle pairs of tail-feathers *buff*. Total length, 23 inches; wing, 10·5; tail, 8·4; tarsus, 4·4.

Adult Female.—Differs from the *female* of *L. rufa* in having the ground-colour of the upper-parts *darker chestnut than the neck*, and the tail *black*. Total length, 22 inches; wing, 9·8; tail, 7·6; tarsus, 3·6.

Range.—Forests of Borneo.

Habits.—This splendid Fire-Back is a bird of the low country, but nothing further has been recorded of its habits, though there is no reason to believe that they differ from those of its Malayan ally *L. rufa*. It is known in Sarawak by the native name of "Sempidan."

NOTE.—In the National Collection there is a skin of a *male* example of a *Lophura* which was sent by Mr. J. R. Reeves from China, and has evidently been in captivity, some of the flight-feathers of both wings having been cut, and a second perfectly similar example of this bird was recently seen living in the aviary of the late Capt. E. W. Marshall at Marlow. Mr. D. G. Elliot is of opinion that the former specimen is a hybrid between *Lophura rufa* and *L. ignita*, but I can see no reason for this conjecture, and should not be surprised if it were to prove to be a species distinct from either of the species mentioned. The bird in the National Collection resembles the male of *L. ignita*, but is distinguished by having the feathers down the middle of the breast and abdomen entirely black, those on the sides margined and largely mixed with black, only the middle part of some of them being rufous-chestnut, and the middle pair of tail-feathers white.

III. DIARD'S CRESTED FIRE-BACK. LOPHURA DIARDI.

Euplocomus diardi, Bonap. C. R. xliii. p. 415 (1856; *ex* Temminck MS.).

Diardigallus prælatus, Bonap. C. R. xliii. p. 415 (1856); Gould, B. Asia, vii. pl. 21 [male only] (1860).

Diardigallus fasciolatus, Blyth, J. As. Soc. Beng. xxvii. p. 280 (1858).

Euplocomus prælatus, Sclater, List of Phas. p. 6, pl. 6 (1863); id. and Wolf, Zool. Sketches (2), pl. 35 (1867); Elliot, Monogr. Phasian. ii. pl. 24 (1872).

Lophura diardi, Ogilvie-Grant, Cat B. Brit. Mus. xxii. p. 290 (1893).

Adult Male.—Head, throat, and crest black, the latter slightly glossed with steel-blue; neck, mantle, and chest grey, *very finely mottled with black;* wing-coverts with a black white-edged band near the extremity; lower back buff, glossed with gold; rump-feathers black, glossed with purplish-blue, widely margined with dark crimson, shot with bronzy-red; rest of under-parts and tail, *including the middle pairs of feathers, black*, glossed with greenish-blue; naked skin on sides of head and wattles *red*. Total length, 24 inches; wing, 9·8; tail, 13; tarsus, 3·4.

Adult Female.—Differs conspicuously from the female of the other species in having the wing-coverts and scapulars *black, with wide-set buff bands;* the breast and sides of the belly chestnut; and the rest of the under-parts brownish-black, margined with white. Total length, 21 inches; wing, 8·8; tail, 8·3; tarsus, 2·9.

Range.—Shan States, Siam, Cambodia, and Cochin China.

Hybrids between this species and the Lineated Kalij Pheasant (*Gennæus lineatus*) have been bred in the Zoological Society's Gardens, London.

Nothing is known of the habits of this splendid Fire-Back, but it is captured and brought down from the interior to Bangkok, whence it is imported to this country in some numbers, and is by no means an uncommon bird in aviaries.

THE WATTLED PHEASANTS. GENUS LOBIOPHASIS.

Lobiophasis, Sharpe, Ann. Mag. N. H. (4), xiv. p. 373 (1874).

Type, *L. bulweri*, Sharpe.

Tail composed of *thirty-two*[*] feathers in the *male* (twenty-

---

[*] By far the largest number of tail-feathers found in any of the *Phasianidæ*. One of the Eared-Pheasants (*Crossoptilon auritum*) has *twenty-four*, and the smallest number occurs in the Painted Quails (*Excalfactoria*), which have only *eight*.

PLATE XX.

BULWER'S WATTLED PHEASANT.

eight in the *female*), compressed and pointed; the middle pairs being very much curved and more than twice as long as the outer pairs, which have little or no web. In all the feathers, the shaft extends considerably beyond the web, and in the outer pairs it terminates in a sharp point.

The first flight-feather is much shorter than the second, which is about equal to the tenth; the fifth is somewhat the longest.

In the *male* the head is almost entirely naked, with the exception of a few feathers down the middle of the crown, and it is ornamented with three pairs of wattles; a large pair, one on each side of the head, a very large one on each side of the throat, and a small pair at the base of the upper mandible.

The feet in the *male* are armed with a pair of short stout spurs.

The plumage of the sexes is quite different. Only one species is known.

1. BULWER'S WATTLED PHEASANT. LOBIOPHASIS BULWERI.

Lobiophasis bulweri, Sharpe, Ann. Mag. N. H. (4), xiv. p. 373 (1874); Gould, B. Asia, vii. pl. 13 (1875); Sclater, P. Z. S. 1876, p. 465, pl. xliv.; Ogilvie-Grant, Cat. B. Brit. Mus. xxii. p. 292 (1893).

Lobiophasis castaneicaudatus, Sharpe, P. Z. S. 1877, p. 94; Gould, B. Asia, vii. pl. 12 (1877).

(*Plate XX.*)

Adult Male.—Neck and chest dark crimson; rest of plumage black, each feather margined with steel-blue; *upper tail-coverts and tail pure white;* bill horn-colour; naked skin of head and wattles bright blue; feet and toes red. Total length, 35 inches; wing, 10·3; tail, 18; tarsus, 3·5.

Immature Male (*L. castaneicaudatus*).—Differs from the *adult* in having the top of the head, chin, and throat thickly covered with purplish-black feathers mixed with rufous, the blue wattles but slightly developed; the dark crimson on the neck and chest much brighter, and the *upper tail-coverts and tail chestnut*, and shaped like those of the female.

Adult Female.—Above brownish-buff, inclining to rufous on the wings, and all finely mottled with black; below rufous and similarly mottled; upper tail-coverts and tail chestnut, with some fine black markings. Total length, 20 inches; wing, 9·4; tail, 6·4; tarsus, 3·7.

Range.—Mountain forests of Sarawak, Northern Borneo.

The male of this magnificent Pheasant, with its curious wattled head and many-feathered pure white tail, is strikingly different from all the other birds of its kind. It was first obtained by Sir Hugh Low, who gave the specimen to Governor Sir Henry Bulwer (after whom it was named), in the mountains bordering the Lawas River in 1874, and since that date a number of specimens have been sent to Europe, but good skins, with perfect tail-feathers, are difficult to obtain, and still command a high price in this country.

A few years after Dr. Sharpe described the first examples, he received from the same locality male and female specimens, which, in the opinion of Mr. Sclater and the late Mr. Gould, represented a second and perfectly distinct species of *Lobiophasis*. This male had the tail chestnut, comparatively short, and much like that of the female of *L. bulweri*. Acting against his own better judgment, Dr. Sharpe described this bird under the name of *L. castaneicaudatus*, but shortly after another male example arrived, in which the chestnut tail-feathers were being replaced by the white feathers of *L. bulweri*, clearly showing that the chestnut-tailed bird is merely the immature of the white-tailed form. We are informed that the perfect white tail is not assumed till the male is in his *third* year, but this requires confirmation, and it appears to me more probable that the full plumage is assumed in the second year.

Habits.—Very little is known about the Wattled Pheasant, for it frequents the dense mountain forests, is extremely shy, and very rarely seen, all the specimens obtained being caught by means of snares.

Mr. C. Hose writes: "Bulwer's Pheasant is only found on the mountains, though it does not ascend very high, not extending beyond 2,000 feet as far as I know. The actions of this bird are entirely Fowl-like, and it is much more like a Jungle Fowl in its ways than a Pheasant. Wolf's picture in the 'Birds of Asia' gives a wrong idea of the carriage of the bird, and I very much doubt whether it ever sits up in the way there depicted. On the contrary, it skulks along through the jungle, carrying its tail in a curve like a Fowl. It is often trapped by the natives and is essentially a ground-bird, seldom taking flight, but preferring to run through the jungle to save itself. I believe that it takes quite three years before the full white tail is assumed. Native name 'Bagier.'"

THE EARED-PHEASANTS. GENUS CROSSOPTILON.

Crossoptilon, Hodgson, J. As. Soc. Beng. vii. p. 864 (1838).

Type, *C. tibetanum*, Hodgson.

Tail composed of *twenty to twenty-four* feathers (the number varying in the different species), large, full, and rounded, the middle pair being twice as long as the outer pair. The extremities of the middle pair much curved, the webs long and decomposed.

First flight-feather shorter than the second which is equal to the ninth or tenth; fifth or sixth somewhat the longest.

Sides of the face naked, red, and covered with small papillæ.

Plumage of sexes similar; *ear-coverts much lengthened and forming a long white tuft on each side of the head.*

Feet in *male* armed with short stout spurs.

I. HODGSON'S EARED-PHEASANT. CROSSOPTILON TIBETANUM.

Phasianus (Crossoptilon) tibetanus, Hodgson, J. As. Soc. Beng. vii. p. 864, pl. 46 (1838); id. Ind. Rev. iii. p. 593, pl. (1839).

Crossoptilon tibetanum, Sclater, List of Phasian. p. 6, pl. 4 (1863); Elliot, Monogr. Phasian. i. pl. 14 (1872); David and Oustalet, Ois. Chine, p. 407, pl. 107 (1877); Hume and Marshall, Game Birds of India, i. p. 115, pl. (1878); Ogilvie-Grant, Cat. B. Brit. Mus. xxii. p. 293 (1893).

Crossoptilon auritum, G. R. Gray (*nec* Pallas), Gen. B. iii. p. 495, pl. cxxv. (1845).

Crossoptilon drouynii, Verreaux, N. Arch. Mus. Bull. iv. p. 85, pl. iii. (1868); Elliot, Monogr. Phasian. i. p. xviii. pl. 15 (1872).

Adult Male.—Crown covered with short, soft, curly black feathers; long ear-tufts white, as in all the other species; *whole plumage* above and below *pure white*, shading into grey on the longer wing- and tail-coverts; quills brownish; tail with *twenty* feathers, black, glossed with dark greenish-blue and deep purple towards the extremity.* Total length, 36 inches; wing, 12·4; tail, 18·6; tarsus, 3·9.

Adult Female.—Perfectly similar in plumage, but devoid of spurs.

Range.—Mountains of Western China and Eastern Tibet.

The typical specimen described by Hodgson was brought into Nepal by an envoy who had been to Pekin, but the exact locality where the bird was obtained was never ascertained.

Habits.—This splendid white Pheasant inhabits the pine-forests at elevations varying from 10,000 to 12,000 feet above

* In Hodgson's type the six outer pairs of tail-feathers have an oblong white spot on the outer web running nearly parallel to the shaft, but these markings are not symmetrical on the two sides, and, in all other specimens that we have examined, are entirely absent.

sea-level. It is extremely sociable in its habits, and it is said that forty or fifty may be found, roosting in company, on the pine-trees.

Abbé David informs us that this white *Crossoptilon* is only met with in some of the wooded localities of China, on the high mountains of Western Sze-chuen, in the neighbourhood of Moupin and Ta-tsien-lou, where its existence is protected by the superstitious respect of the natives. It is a very gregarious bird, loving to live in company with many of its kind, even when engaged in rearing its young, and it does not wander far from the place where it is bred. It feeds on leaves, roots, grains, and insects. Fortunately for its safety, the flesh of this Eared-Pheasant is but moderately good to eat, and sportsmen prefer the smaller Pheasants (*Phasianus*) as game, since they are more widely distributed and easier to procure.

II. THE WHITE-TAILED EARED-PHEASANT. CROSSOPTILON LEUCURUM.

Crossoptilon leucurum, Seebohm, Bull. Brit. Orn. Club, No. iv. p. xvii. (1892); Ogilvie-Grant, Cat. B. Brit. Mus. xxii. p. 294 (1893).

Adult Male.—Distinguished from *C. tibetanum* by having *the greater part of the tail-feathers white*, all being pure white, with black extremities glossed with purplish-blue.

Adult Female.—Has the white on the tail-feathers less extensive, and the middle and outer pairs have the inner webs grey, while all are tipped and margined with dark grey.

Range.—Eastern Tibet; met with between the Sok Pass and Chiamdo, also on the plateau between the Sok Pass and Lhassa.

The typical examples of this apparently perfectly distinct species were obtained by Captain Bower and Dr. Thorold between the Sok Pass and Chiamdo, and similar specimens were collected by Prince Henry of Orleans and M. Bonvalot a few years before on the plateau between the Sok Pass and

Lhassa. The latter birds are now in the Paris Museum, and Dr. Oustalet regards them as merely varieties of *C. tibetanum*, or hybrids between this species and the slate-grey *C. auritum*, Pallas, which has the greater part of the outer tail-feathers white. We entirely agree with Mr. Seebohm in believing this conclusion to be a mistake, for *C. auritum*, we may further remark, has the tail composed of twenty-four, *not* twenty feathers.

It appears that the range of *C. leucurum* overlaps that of *C. tibetanum* in Eastern Tibet, and it may be that in this locality the two forms interbreed, so it is just possible that Hodgson's type of *C. tibetanum*, which has some white markings on the six outer pairs of tail-feathers (see previous footnote, p. 252) may be a cross-bred bird of this description, but the exact locality where it was obtained is quite uncertain.

III. THE MANCHURIAN EARED-PHEASANT. CROSSOPTILON MANCHURICUM.

Crossoptilon auritum sive *mantchuricum*, Swinhoe, P. Z. S. 1862, p. 286, and 1863, p. 306.

Crossoptilon auritum, Sclater (*nec* Pallas), List of Phasian. p. 6, pl. 5 (1863); Milne-Edwards, N. Arch. Mus. Bull. i. p. 12, pl. i. figs. 1 and 2 (1865); Gould, B. Asia, vii. p. 22 (1870).

Crossoptilon mantchuricum, Elliot, Monogr. Phasian. i. pl. 16 (1872); David and Oustalet, Ois. Chine, p. 405, pl. 106 (1877); Sclater, P. Z. S. 1879, p. 118, pl. viii. fig. 5; Ogilvie-Grant, Cat. B. Brit. Mus. p. 294 (1893).

Adult Male.—Differs chiefly from *C. tibetanum* in having an indistinct white band across the crown, the neck black, shading into brown on the mantle, the lower back and rump dirty white; chest blackish-brown; rest of under-parts lighter. Tail with *twenty-two* feathers, the basal part dirty white and the ends brownish, glossed with rich purplish-blue. Total length, 40 inches; wing, 12·7; tail, 22·6; tarsus, 4·1.

Adult Female.—Differs only in having no spurs.

Range.—Mountains of Manchuria and Pechi-li.

Swinhoe says: "This bird is called *Ho-ke* by the natives. The character *Ho* is a peculiar one, and especially applied to this bird from ancient times. It does *not* mean *Fire*, as Mr. Saurin states in his account of the bird in the 'Proceedings of the Zoological Society.' *Ke* means *Fowl*. The feathers of this bird were formerly worn by Tartar warriors."

According to Abbé David, the brown *Crossoptilon*, which is known by the name of *Hoky* in Pekin, is resident on some of the wooded parts of the mountains of Pechi-li, but for some years past it has become very rare, and it cannot be long before it completely disappears, partly on account of the constant persecution it is subjected to, and partly from the destruction of the woods which form its headquarters. It is an extremely gentle and sociable bird, living in large flocks, and subsisting chiefly on grain, buds, leaves, roots, and insects. It seems well adapted for domestication, the more so as it is easily fed; but in captivity one must provide the shade of a park and the neighbourhood of a clear stream of water—that is, similar surroundings to those it is accustomed to in its wild state.

Mr. Misselbrook writes: "Hens lay from twelve to sixteen eggs each at a setting, the time of incubation being about twenty-eight or thirty days." This refers, of course, to birds in captivity.

Eggs.—Uniform pale stone-colour. Measurements, 2·3 by 1·7 inches.

IV. PALLAS' EARED-PHEASANT. CROSSOPTILON AURITUM.

Phasianus auritus, Pallas, Zoogr. Rosso-Asiat. ii. p. 86 (1811).
Crossoptilon auritum, Elliot, Monogr. Phasian. i. pl. 17 (1872):
    Prjevalsky, in Rowley's Orn. Misc. ii. p. 420 (1877):
    David and Oustalet, Ois. Chine, p. 406, pl. 108 (1877);
    Ogilvie-Grant, Cat. B. Brit. Mus. xxii. p. 295 (1893).
Crossoptilon cærulescens, David, MS.; Milne-Edwards, C. R.
    lxx. p. 538 (1870).

Adult Male.—General colour slate-grey; an indistinct white band bordering the black crown behind; chin and throat white; tail composed of *twenty-four* feathers, the six outer pairs with the basal three-quarters white and the ends black, glossed with purple. Total length, 40 inches; wing, 12·4; tail, 21; tarsus, 4.

Adult Female.—Like the *male*, but devoid of spurs.

Range.—The mountains of Koko-nor, Kansu, and North-western Sze-chuen, Western China.

Habits.—This Pheasant inhabits the wooded mountainous regions, and ascends to a height of even 10,000 feet above the sea-level. According to Prjevalsky, "it is a resident and remains all the year round in certain places. Water does not seem to be of so much necessity to this bird as it is to other species of the present group; at least, it keeps very often to localities in the Ala-Shan Mountains, where not a drop of water is to be found.

"In autumn and winter they congregate in small flocks, probably in families, but very early in spring separate into pairs, when the males at once commence to crow—*i.e.*, uttering at intervals a loud disagreeable note somewhat resembling the cry of a Peacock. This usually occurs in the morning, but occasionally also during the day. . . .

"After the breeding season the males at once commence moulting, and attain their fresh plumage only in October again. Generally their feathers very soon get worn, and the birds are in full plumage only for a short time in winter and spring.

"Like most of the Family, these birds are fond of digging about in the ground in search of roots; and it appears that they chiefly feed upon plants."

Eggs.—Vary from five to seven; smooth, pale olive-grey in colour, without any spots, and much like those of the Common Fowl. Measurements, 2·16 by 1·6 1·63 inches.

V. HARMAN'S EARED PHEASANT. CROSSOPTILON HARMANI.

Crossoptilon harmani, Elwes, Ibis, 1881, p. 399, pl. xiii.;
  Ogilvie-Grant, Cat. B. Brit. Mus. xxii. p. 296 (1893).

Adult Male.—Like *C. auritum*, but distinguished by having a *wide and well-marked white band* bordering the back of the head between the ear-coverts, and no white on the basal part of the outer tail-feathers. The full number of tail-feathers is *probably twenty-four*, as in *C. auritum*, for though the unique type-specimen, which is in bad condition, has only nineteen feathers remaining, the two middle pairs appear to be entirely wanting, as well as one on the left side.

Range.—Tibet, 150 miles east of Lhassa.

Remarks.—Mr. H. J. Elwes says: "For this fine species I am indebted to Lieut. Harman, R.E., who has displayed himself as a surveyor and explorer of the Eastern Himalayas, especially in Sikhim, where he has been employed for some years. When at Darjeeling in December last, I saw the skin of what I at once recognised as a new *Crossoptilon* hanging on the wall of his room. Unfortunately it had never been properly preserved, and was in such a terribly moth-eaten state that the remains, which he kindly presented to me, and which are now in the British Museum, are hardly worth preserving. They have, however, proved sufficient for Mr. Keulemans to make a very accurate drawing, the only fault of which is that the ear-coverts do not seem in the specimen to be so strongly developed as in the figure.

"The skin was brought to Mr. Harman by one of his native surveyors, who said that he had procured it 150 miles east of Lhassa, at an elevation of about 6,000 feet, where it was found in flocks during winter. This part of Thibet has never been visited by any European, or by any of the late Mr. Mandelli's native hunters, and having, as reported, a much milder climate and more luxuriant vegetation than the western parts of Thibet, may be expected to produce a number of remarkable and, as yet, unknown species."

THE KALIJ PHEASANTS. GENUS GENNÆUS.

Gennæus, Wagler, Isis, 1832, p. 1228.

Type, *G. nycthemerus* (Linn.).

Tail composed of *sixteen* feathers, long, laterally compressed (like that of the Game-Cock); the middle pair somewhat, or considerably, longer than the second pair, and at least three times the length of the outer pair in the *male*.

First flight-feather considerably shorter than the second, which is equal to the ninth or tenth; fifth or sixth somewhat the longest.

Sides of the head naked. Plumage of sexes quite different.

Male with a long hairy crest and armed with a pair of stout, fairly long, spurs.

I. THE WHITE-CRESTED KALIJ PHEASANT. GENNÆUS ALBOCRISTATUS.

Phasianus albocristatus, Vig. P. Z. S. 1830, p. 9, and 1832, p. 16;
Gould, Cent. B. Himal. pls. 66, 67 (1832).

Phasianus hamiltoni, J. E. Gray, Ill. Ind. Zool. i. pl. 41 (1830-32).

Euplocamus albocristatus, Elliot, Monogr. Phasian. ii. pl. 18 (1872); Hume and Marshall, Game Birds of India, i. p. 177, pl. (1878); Oates, ed. Hume's Nests and Eggs Ind. B. iii. p. 413 (1890).

Gallophasis albocristatus, Mitch. P. Z. S. 1858, p. 544, pl. 148, fig. 1, and pl. 149, fig. 3.

Gennæus albocristatus, Wagler; Ogilvie-Grant, Cat. B. Brit. Mus. xxii. p. 298 (1893).

Adult Male.—A long crest of hairy *white* feathers; rest of the upper-parts and throat black, glossed with purplish and steel-blue; the *mantle* and upper tail-coverts narrowly bordered with dirty white, and the lower back and rump more widely margined with pure white; fore-neck and chest dirty white, shading into whitish-brown on the rest of the under-parts, all the feathers of which are long and pointed. Total length, 25 inches; wing, 9·3; tail, 11; tarsus, 3.

Adult Female.—A long *brownish-grey* hairy crest; general colour of the rest of the plumage reddish-brown, brighter on the rump and under-parts; the upper-parts finely mottled with black and edged with grey, the wing-coverts and under-parts with white; under-parts with *white shafts*, never pale shaft-stripes; throat and middle of belly dirty white; outer tail-feathers black. Total length, 22·5 inches; wing, 8·8; tail, 8·8; tarsus, 2·5.

Range.—The lower and middle ranges of the Western Himalayas, from Hazara to Nepal and Western Kumaon.

Habits.—According to Mr. Hume the White-crested Kalij is found "throughout the fairly wooded lower and middle ranges of the Himalayas, from Kumaun to Hazára, here sparingly, there abundantly, according to season and a variety of other more or less potential influences."

The late Mr. Frederic Wilson says: "This well-known Kalij is most abundant in the lower regions; it is common in the Dhún at the foot of the hills, in all the lower valleys, and everywhere to an elevation of about 8,000 feet: from this it becomes more rare, though a few are found still higher. . .

"In the lower regions it is found in every description of forest, from the foot to the summit of the hills; but it is most partial to low coppice and jungle, and wooded ravines or hollows. In the interior it frequents the scattered jungle at the borders of the dense forests, thickets near old deserted patches of cultivation, old cowsheds and the like, coppices near villages and roads, and, in fact, forests and jungle of every kind, except the distant and remoter woods, in which it is seldom found. The presence of man, or some trace that he has once been a dweller in the spot, seems, as it were, necessary to its existence. . . . .

"Their call is a loud whistling chuckle or chirrup; it may occasionally be heard from the midst of some thicket or coppice at any hour of the day, but is not of very frequent occurrence. It is generally uttered when the bird rises, and, if it flies into a tree near, is often continued some time. When flushed by a cat or a small animal, this chuckling is always loud and earnest. The Kalij is very pugnacious, and the males have frequent battles. On one occasion I had shot a male, which lay fluttering on the ground in its death struggles, when another rushed out of the jungle and attacked it with the greatest fury, though I was standing reloading the gun close by. The male often makes a singular drumming noise with its wings, not unlike the sound produced by shaking in the air a stiff piece of cloth. It is heard only in the pairing-season; but whether to attract the attention of the females or in defiance of his fellows, I cannot say, as I have never seen the bird in the act, though often led to the spot where they were by the sound."

Mr. Hume remarks, however, "This is certainly not to attract the females, but solely as a defiance. If you peg out a tame male of the allied Vermicellated Pheasant in the breeding-season, as is commonly done in Burma, surrounding him with snares, and then set your male drumming, by imitating the sound with a piece of stiff cloth, male after male replies, rushes in at your bird and gets caught in the snares, but no female ever puts in an appearance, or is ever thus snared."

According to Mr. Wilson, the species feeds on roots, grubs, insects, seeds and berries, and the leaves and shoots of shrubs.

The following remarks of Captain J. H. Baldwin are worth quoting. He says: "I have flushed this Pheasant and the common Red Jungle Fowl from the same description of cover at the foot of the hills. The call of the bird, which may be heard at all times of the day, is a sharp *twut, twut, twut*, sometimes very low, with a long pause between each note, then suddenly increasing loudly and excitedly. Generally speaking, when uttering this cry, which at times might be mistaken by anyone unacquainted with it, for that of some small bird, the Kalij is alarmed by a prowling Marten or Hawk hovering overhead, perhaps a dog, but still oftener it is heard when a pair of cocks are about to engage in mortal combat.

" Not unfrequently a cunning old cock, instead of taking wing at once when the dog is close upon him, has a provoking habit, most irritating to both dog and master, of flying up into a tree, making a prodigious clucking the while, and at the same time taking a look round to see if the coast is clear. The bird in this manner often observes where the gun is posted, and then takes wing in a safe direction.

"The Kalij Pheasant, when alarmed, will generally fly down the *Khad*, and will often take along the side of the hill. Though it will *run*, yet it will hardly ever *fly* up hill. Its speed when well on the wing is amazing, greater frequently, I am certain, that any rocketer out of an English cover. When not bullied by the hill-men, they will come close up to the backs of villages, especially if there are fields of corn at hand. I have shot them out of standing crops when the fields are situated near the jungle."

Referring to the whirring sound they make most commonly, but not exclusively, in the breeding-season, he says :

"We had been sitting motionless for, I suppose, half an hour, when I was startled, all of a sudden, by the loud drumming noise I have already described, close at hand. The sound came from behind, and on looking over my shoulder, my companion with a smile pointed out the drummer. An old cock Kalij was squatting on the stump of a fallen tree, and with its feathers all ruffled and tail spread, was causing this extraordinary sound by rapidly beating its wings against its body."

Nest.—Generally placed on the ground under a rock or bush and composed of a few dead leaves and grass.

Eggs.—Usually eight in number; varying in colour from creamy-white to reddish-buff, the shell glossy and finely pitted with minute pores. Average measurements, 1·94 by 1·44 inch.

II. NEPAL KALIJ PHEASANT.   GENNÆUS LEUCOMELANUS.

Phasianus leucomelanos, Lath. Ind. Orn. ii. p. 633 (1790).
Euplocamus leucomelanus, Hume and Marshall, Game Birds of India, i. p. 185. pl. (1878).
Gallophasis leucomelanus, Scully, Str. F. viii. p. 345 (1879).
Gennæus leucomelanus, Ogilvie-Grant, Cat. B. Brit. Mus. xxii. p. 300 (1893).

Adult Male.—Like the male of *G. albocristatus*, but the crest is *black*, and the terminal bars to the feathers of the rump and upper tail-coverts are usually narrower. Also rather smaller in size.

Adult Female.—Like the female of *G. albocristatus*, but rather darker, especially on the under-parts, which are dark reddish-brown or dark brown.

Range.—Mountain forests of Nepal, to an elevation of about 9,000 feet above sea-level.

The habits and nidification of this species are, of course, very similar to those of the other Kalij Pheasants.

Dr. Scully says: "*G. leucomelanus* is common wherever thick forest is found, from Hitorna in the Nepal Dún to the Valley of Nepal; in all the wooded hills surrounding the latter, up to an elevation of nearly 9,000 feet; and in every forest about Noakote. It is usually seen in pairs or in parties of from three to ten, often feeding on the ground near cultivated patches at the borders of forest.

"The birds seem very fond of perching on trees, and it is usually in this position that one comes across them in forcing one's way through forest which has a dense undergrowth. On such occasions the Kalij first gives notice of its whereabouts by whirring down with great velocity from its perch, and then running rapidly out of sight to the shelter of some thicket. In the winter the birds roost on trees at the foot of the hills, and the plan for making a bag is to post oneself, about sunset, under some trees which they are known to frequent, and await their coming. The birds are then soon heard threading their way through the jungle towards their favourite trees, and at once fly up and perch. When once settled for the night in this way, they are not easily alarmed, and I have shot four or five birds in quick succession before the rest of the party would clear out to quieter quarters. Occasionally, too, one can get a shot at the Kalij as they cross a hill-path through the forest, on their way to or from some stream.

"Great numbers of the Nepal Kalij are snared and brought into Khatmandu for sale. The birds bear confinement in the valley very well, and I reared several chicks to maturity."

Nest and Eggs.—Very similar to those of *G. albocristatus*.

III. THE BLACK-BACKED KALIJ PHEASANT. GENNÆUS MELANONOTUS.*

? *Phasianus muthura*, Gray, in Griff. ed. Cuv. iii. p. 27 (1829).
Euplocamus melanotus (Blyth), Hutton, J. As. Soc. Beng. xvii. pt. 2, p. 694 (1848); Elliot, Monogr. Phasian. ii. pl. 19 (1872).
Gallophasis melanotus, Mitch. P. Z. S. 1858, p. 544, pl. 149, fig. 2.
Euplocomus melanonotus, Hume and Marshall, Game Birds of India, i. p. 191, pl. (1878); Oates, ed. Hume's Nests and Eggs, iii. p. 415 (1890).

* Though I still have little doubt that Latham's "Chittygong Pheasant," on which Gray founded his *Phasianus muthura*, refers to the present species, Mr. W. T. Blanford has recently called my attention to the fact that Latham describes his bird as being as big as a Turkey, Gray of course following suit. As there is thus some doubt as to the propriety of using the name of *G. muthura* for this bird, I have thought it better to use the much more appropriate and descriptive name of *G. melanonotus* (Blyth).

Gennæus muthura, Ogilvie-Grant, Cat. B. Brit. Mus. xxii. p 301 (1893).

Adult Male.—Differs from the male of *G. albocristatus* in having the crest *black*, the upper-parts with a brighter purplish gloss, and the feathers of the lower back and rump glossed with deep purplish-blue and *without* white terminal bands.

Adult Female.—Quite similar in plumage to the *female* of *G. leucomelanus*.

Range.—Forests of Sikhim, Native Sikhim, and Western Bootan. Perhaps found in Eastern Nepal.

Habits.—Mr. Gammie furnishes the following excellent account of this species:—

"In Sikhim the Black-backed Kalij is abundant from about 1,000 up to 6,000 feet, and it is occasionally found at both lower and higher elevations. It frequents forest and scrub, rarely coming out to cleared land, except in the mornings and evenings to feed, and even then seldom leaving the cover for many yards.

"At no time of the day is it a shy bird, but in the evenings and early mornings it is almost as tame as a domestic fowl, and, if feeding on the road, will leasurely walk but a few steps out of the way of a passer-by.

"It appears to dislike sunshine, and scarcely leaves the shade of trees or shrubs while the sun is up.

"It seldom, if ever, perches in the daytime, but keeps to the ground, unless suddenly disturbed by dogs or wild animals, when it may take refuge in a tree as a last resource. If alarmed by men it always runs along under the scrub if the circumstances are favourable for that mode of escape; but if not, it flies within twenty feet of the ground for forty or fifty yards, and then again alights on the ground. By making a short detour they will be found close to where they alighted.

"Usually it is a silent bird, but when suddenly alarmed it utters a sharply repeated '*koorchi, koorchi, koorchi,*' as it rises on the wing. When, however, the males are in the fighting humour—which they usually are about breeding-time—their call, as they advance towards each other, is '*koor koor, waak waak*'; the former being the threatening, and the latter the attacking note. They also at times answer each other's calls in the jungles.

"In fine weather the male often makes a sharp drumming noise by beating his wings against his sides, somewhat after the style of the wing-flapping of a domestic cock preparatory to crowing from some elevated place; but instead of the cock's few leisurely flaps, the Kalij strikes oftener and smarter, producing a sound more like drumming than flapping. . . .

"The natives look on the drumming of the Kalij as a sure sign of approaching rain. It is heard at all seasons of the year, but most frequently before the setting in of the rainy season; at other times generally just before a fall of rain.

"The food of the Kalij is varied in the extreme. It eats almost everything in the shape of seeds, fruit, and insects, but is particularly fond of the larvæ of beetles out of cow-dung and decayed wood, and of several of the jungle yams which bear tubers along their vines at the axils of the leaves. When the vine-borne tubers are exhausted, it will scratch away the soil to get at those underground."

Nest and Eggs.—Similar to those of *G. albocristatus*. The average measurement of the latter is 1·91 by 1·47 inch.

The three Himalayan species of Kalij which I have just dealt with are very easily distinguished one from another, and so far as I know do *not* intergrade, though it is possible that where the range of *G. leucomelanus* touches or overlaps (if it does either) the habitats of *G. albocristatus* and *G. melanonotus*, respectively to the west and east, intermediate forms may occur. When we consider the Burmese Pheasants, however, the different forms of Kalij are by no means so easily dealt with; for, though there are three well-marked principal forms, absolutely distinct from one another and occupying widely different geographical areas, there can be no doubt that, given a large enough series of specimens from the intervening country between the headquarters of any two, a chain of intermediate forms would be found, and the extreme types would be shown to grade imperceptibly into one another. For the sake of clearness the accompanying sketch-map of Burma has been prepared, showing the countries in which each form occurs, the range of the three principal forms being shown by the shaded areas marked A, B, and C, while the places where the intermediate birds are known to occur are marked AB, BA, &c., thus indicating their affinities to one or other of the main types. Thus *G. cuvieri* is marked AB, which shows that it is most nearly allied to *G. horsfieldi* (A), and less nearly related to *G. lineatus* (B); while in *G. oatesi*, indicated by the letters BA, the reverse obtains.

The Black-breasted Kalij (A) ranges over a large part of Northern Burma, extending in the north to Eastern Bootan, in the west to Chittagong, south to the Northern Arakan Hills, and east as far as Bhâmo.

The Vermicellated Kalij (B) is met with in Pegu as far west as the valley of the Irrawady, in Northern Tenasserim and North-western Siam, and, according to Oates, extends up the Irrawady Valley as far as Bhâmo, though this latter statement requires confirmation.

Anderson's Kalij (C), first obtained in the Kachin Hills east of Bhâmo, has since been met with at Dargwin, and is probably found from that place northwards along the Salween Valley to Yun-nan, but its range is not yet defined.

It must first be stated that A does not intergrade with the Black-backed Kalij of Sikhim and Western Bootan, and *if* the ranges of the two birds do overlap in Central Bootan, which they probably do not, no intermediate birds have as yet been recorded.

The Northern Arakan Hills is the most southern point

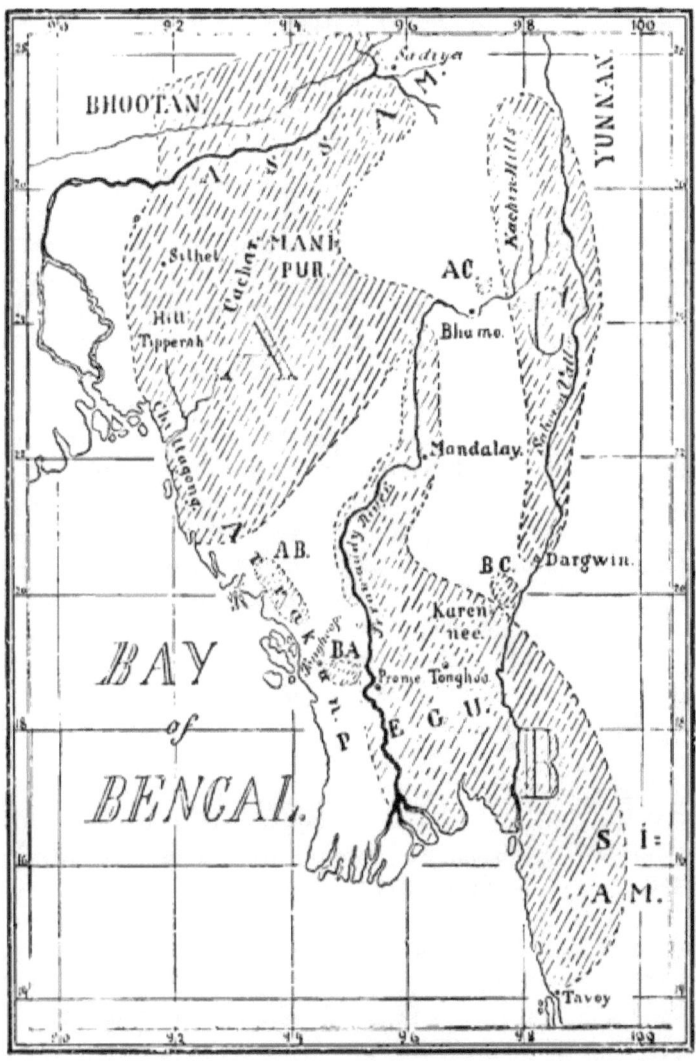

Sketch-map of the Burmese countries, shewing the ranges of the species and races of the Kalij Pheasants.

where A is found, and as we go southwards along this range we meet with the bird known as *G. cuvieri* (AB), very nearly allied to A, but the male has the whole of the black upperparts finely and irregularly pencilled with white lines, and the female is also somewhat different. In the south of the Arakan range, on the road between Prome and Tonghoo, *G. oatesi* (BA) occurs, the males being evidently much more nearly allied to B than to A, though they resemble A in having the feathers of the lower back and rump more or less distinctly margined with white. The female differs from all the allied forms in having the outer tail-feathers mostly chestnut.

It will thus be seen that in Arakan, a large tract, between the ranges of A and B, we find two intermediate forms of Kalij, practically bridging over the great differences between typical specimens of A and B, and no doubt with a large series of birds from all parts of Arakan every intermediate stage could be found, those from the north gradually merging into A and those from the south into B.

Again, between A and C we have a male example, the type of *G. davisoni* (AC) from the Kachin Hills, just to the east of Bhâmo, which is perfectly intermediate in plumage between A and C. The female of this form is still unknown.

Lastly, between B and C more or less intermediate birds are to be found in the neighbourhood of Karen-nee (BC), south of Dargwin, but they are so nearly allied to C, that we have thought it unnecessary to call them by a distinct name.

It will thus be seen that all the three forms, A, B, and C, which are so perfectly distinct *inter se*, have connecting links, which are met with in the intermediate districts joining their various ranges, where typical examples of A, B, and C are *not* to be found.

It is always a difficult matter to deal with such intermediate forms as those we have just described. It is quite wrong to apply the word hybrid to them, for they are really incipient species, occupying a tract of country where neither typical A, B, nor C are met with. There will always be found some people who disapprove of calling these intermediate forms by distinct names, but after all it is only a matter of convenience, and perhaps the most satisfactory plan is that which we have followed—viz., treating them as sub-species of the type to which they show most affinity.

Another way of getting out of the dilemma is by the use of trinomial nomenclature. To those who follow this objectionable plan, *G. cuvieri* would be known as *G. horsfieldi lineatus*, and *G. oatesi* as *G. lineatus horsfieldi*.

Lastly, I must remind my readers that the *female* of *G. oatesi* (BA) is somewhat different in plumage from both B and A, and may be easily distinguished from all the females of the various species of *Gennæus*.

I have at present only been able to examine a very limited number of skins of these intermediate forms, as few Europeans have visited the countries where they occur, but it is greatly to be hoped that those who have opportunities of visiting the Arakan Hills or Upper Salween Valley at some future time, will endeavour to shoot and preserve all the Kalij they come across. A good series of these intermediate forms would be an extremely valuable and welcome addition to the National Collection.

IV. THE BLACK-BREASTED KALIJ. GENNÆUS HORSFIELDI.

Gallophasis horsfieldi, G. R. Gray, Gen. B. iii. p. 498, pl. cxxvi. (1845); Mitchell, P. Z. S. 1858, p. 544, pl. 148, fig. 2, pl. 149, fig. 1; Sclater, in Wolf's Zool. Sketches (2), pl. 39 (1861).

Euplocamus horsfieldi, Elliot, Monogr. Phasian. ii. pl. 20 (1872); Hume and Marshall, Game Birds of India, i. p. 198, pl. (1878); Oates, ed. Hume's Nests and Eggs, iii. p. 416 (1890).

Gennæus horsfieldi, Ogilvie-Grant, Cat. B. Brit. Mus. xxii. p. 302 (1893).

Adult Male.—Entire plumage black, glossed with purplish or steel-blue; only the feathers of the lower back, rump, and upper tail-coverts margined with white, those of the under-parts being only slightly pointed. Size the same as in the other species.

Adult Female.—Like the *female* of the last named species, but the feathers of the under-parts usually have narrow buff shaft-stripes, and in old examples the middle pair of tail-feathers become uniform dark chestnut, usually contrasting rather strongly with the olive-brown rump; the outer pairs black.

Range.—The forests of Eastern Bootan, Assam, Sylhet, Cachar, Manipur, Hill Tipperah, Chittagong, and North Arakan.

Mr. Hume tells us that "the range of this species is decidedly lower than that of either of the other three; it is common down in the low country along the edges of cultivation and the banks of rivers where there is forest, only a few hundred feet above sea-level, but it grows less plentiful, I am assured, as you ascend the hills, and is very rarely shot at elevations exceeding 4,000 feet."

Mr. R. A. Clark, of the Mynadhar Tea Garden in Cachar, says: "These birds are very common here, keeping to well-wooded hills and ravines. They go about in pairs, though parties of three and four are often met with, and on one occasion I saw a party of eleven. . . . .

"I once witnessed a fight between a male Kalij and a Jungle Cock (*Gallus gallus*) for the possession of a white-ant hill from which the winged termites were issuing. I watched the contest for a quarter of an hour, by which time both birds were exhausted, when the Kalij fled, leaving the Jungle Cock in possession. On another occasion I came across a pair of male Kalij fighting amongst a lot of ferns; they were so taken up with their own affairs that they did not notice my having approached to within fifteen yards; I let them go on for ten minutes, and then went up and caught both; they were quite exhausted; the feathers from the head and neck had all been knocked off, and the latter was bleeding in both birds."

Mr. Cripps writes: "The northern part of the district of Sylhet is covered with low 'teelahs,' or hillocks, between which run small brooks, the whole being overgrown with dense tree-, bamboo-, and cane-jungle, forming dark, damp retreats, such as are the favourite resorts of this species.

"Here they scratch about amongst the fallen leaves for insects, and towards evening and in the early morning stray into any adjacent patches of cultivation, or are to be found feeding about the roadside where these lie within the forests."

Nest.—A heap of dry leaves, with rather a deep cavity scratched in the middle, placed at the foot of a tree.

Eggs.—Like those of *G. albocristatus*, &c., and of the usual Kalij type. Varying in colour from pale buff to rich brownish-buff. Average measurements, 1·85 by 1·48 inch.

SUB-SP. *a*. CUVIER'S KALIJ PHEASANT. GENNÆUS CUVIERI.

Lophophorus cuvieri, Temm. Pl. Col. v. pl. 10 [No. 1] (1820). *Gennæus cuvieri*, Ogilvie-Grant, Cat. B. Brit. Mus. xxii. p. 303 (1893).

Adult Male.—Like the *male* of *G. horsfieldi*, but all the upper-parts are finely pencilled with irregular wavy white lines.

Adult Female.—Like the *female* of *G. horsfieldi*, but all the tail-feathers are more or less mixed with dull rufous, mottled with black, the outer pairs only being black towards the tips.

Range.—The middle and northern Arakan Hills, extending into Chittagong.

SUB-SP. *b*. DAVISON'S KALIJ PHEASANT. GENNÆUS DAVISONI.

Gennæus davisoni, Ogilvie-Grant, Cat. B. Brit. Mus. xxii. p. 304 (1893).

A somewhat Immature Male.—Differs from *G. cuvieri* in having the white lines on the black feathers of the upper-parts, especially the mantle, coarser and more regular, and running more or less parallel to the margins of the feathers. The plumage, in fact, is not unlike that of *G. andersoni*, but the white lines are narrower and the black interspaces broader than in the latter.

Range.—Kachin Hills, east of Bhâmo.

Only one male example of this form is known, and the female has still to be obtained, but will *probably* be found to be intermediate in plumage between *G. horsfieldi* and *G. andersoni*, with the white shaft-stripes on the breast-feathers rather wide, and the outer tail more or less barred with white.

V. THE VERMICELLATED KALIJ PHEASANT. GENNÆUS LINEATUS.

Phasianus lineatus, Vig. Phil. Mag. 1831, p. 147; Jardine and Selby, Ill. Orn. new series, pl. 12 (1836).

Phasianus reynaudii, Less. Bélang. Voy. Ind. Orient. p. 276, pls. 8, 9 (1834).

Gennæus lineatus, Ogilvie-Grant, Cat. B. Brit. Mus. xxii. p. 304 (1893).

Phasianus fasciatus, McClell. Calcutta Journ. N. H. ii. p. 146 pl. iii. (1842).

Euplocamus lineatus, Sclater and Wolf, Zool. Sketches (2), pl. 38 (1861); Elliot, Monogr. Phasian. ii. pl. 23 (1872); Gould, B. Asia, vii. pl. 14 (1875); Hume and Marshall, Game Birds of India, i. p. 205, pl. (1878); Oates, ed. Hume's Nests and Eggs Ind. B. iii. p. 416 (1890).

Adult Male.—Upper-parts finely vermicellated with alternate black and white lines running mostly across the feathers; long crest and under-parts black, with some bluish gloss, the feathers bordering the breast and belly with white shaft-stripes.* Total length, 29 inches; wing, 9·7; tail, 12; tarsus, 3·3.

Adult Female.—Upper-parts olive-brown with V-shaped white marks on the back; crest tinged with rufous; outer webs of the secondary quills mottled with buff and black *along the margin only; sides of the neck with triangular white spots;* throat and fore-neck whitish; under-parts brownish-chestnut, each feather with a pointed white shaft-stripe; middle pair of tail-feathers buff, mottled with black, the outer pairs reddish-brown, with wide irregular white bars edged with black. Total length, 22·5 inches; wing, 8·9; tail, 8·7; tarsus, 2·7.

Range.—From the Irrawady Valley eastwards through the Pegu, Tonghoo, and Karen Hills; extending southwards into Tenasserim as far south as Tavoy, and eastwards into North-western Siam. It is also said to extend northwards along the Irrawady Valley as far as Bhâmo.

Habits.—Mr. Hume informs us: "It is not a bird of high elevations; I have no record of its having been seen even as high as 4,500 feet; it appears to be most numerous at from 1,000 to 3,000 feet, though it certainly occurs as high as 3,500, and again right down to sea-level.

"Its home appears to be the thin deciduous-leaved woods, especially those much mingled with bamboos, of the low hills. It is rarely seen in dense evergreen forests or in grass prairies."

Mr. Oates remarks: "This species is common throughout the whole of Pegu east of the Irrawady.

"It is rare or common just in proportion as the country is level or mountainous. In the plains or undulating portion of Upper Pegu it will be met with in small numbers, if the ravines and nallas are sufficiently precipitous to suit its taste; but in these places, at the best, only one or two will be shot in a long morning's work. It is not till we get to the foot of the hills that this Pheasant can be said to be common. Here the nallas,

---

* In some perfectly adult birds nearly all the feathers of the breast and belly have white shaft-stripes, but this character is apparently individual.

with their pools of water and rocky beds, are particularly favourable to it. As we mount higher, it increases in numbers to such an extent that it is no difficult matter to knock over half a dozen in a morning while marching, and that without leaving the path.

"This Pheasant is averse to all cultivation, and shuns even the *yahs*, or hill gardens, of the Karens, though these may be several miles from the nearest *tay*, or village. It must have thick cover, even while feeding. In the mornings it comes out to feed on the ridges, where the jungle is a trifle less thick than in the valleys. At nine or ten o'clock it descends into the valleys, and after drinking retires into some small secondary watercourse for its mid-day siesta. At this period of the day seven or eight may be found together if it is not the breeding-season. When feeding, they go singly or in pairs. Their food is very varied. Ants, both white and black, are eagerly sought after; the former are an especial weakness of our bird, and the only food on which it thrives in captivity. During the hot weather Pheasants eat the fig of the Peepul ravenously; and I have shot birds with nothing but this food in the stomach.

"The breeding-season begins about the 1st March, and by the end of the month all the hens have commenced laying. It is during this month only that the male makes that curious noise with his wings which seems peculiar to the Kalij group. It may be imitated very fairly by holding a pocket-handkerchief by two opposite corners and extending the arms with a jerk. This noise, made only by the male, is undoubtedly a challenge to other cocks. I have frequently hidden myself near a bird thus engaged, and on two occasions shot cock birds running with great excitement towards the sound.

"The chickens, as soon as they are hatched, are very strong on their legs, and run with great speed. I was fortunate enough to capture portions of four broods. It is astonishing in what a short time the little birds make themselves invisible.

It is difficult to secure more than two out of one batch. It is a case of pouncing on them at once or losing them. The mother is a great coward, running away at the slightest alarm, and thus contrasting very unfavourably with the Jungle Fowl, which keeps running round and round the intruder with great anxiety, till her young ones are in safety. The young ones are very difficult to rear. From some cause or other they become paralysed, lose the use of their legs, languish, and die. This Pheasant is not very shy; on the contrary, it is rather tame; but it has the habit of sneaking quietly away, and very few birds will be seen by one who does not know its peculiarities. It never takes wing unless suddenly surprised, when it will skim across the valley and alight again as soon as possible. Its only call is a low chuckle, frequently uttered both when alarmed and when going to roost."

Writing from Northern Pegu, Captain Feilden says: "An old male is a most extraordinary-looking bird. The tail only is seen moving through the long grass, and I invariably thought at first that it was some new Porcupine or Badger, or some animal. The note, too, adds to the deception; it reminded me a little of the cries of young Ferrets."

It is curious how the habits of this species differ in different parts of its range. The late Mr. W. Davison tells us that in Tenasserim "they come continually into the open to feed about rice-fields and clearings. They are shy, and usually run in preference to flying when disturbed, except when put up by a dog, when they immediately perch. . . . .

"They seem to prefer bamboo, or moderately thin tree-jungle, to dense forest."

Referring to the mode in which the Burmans capture this Kalij by means of a decoy bird, he goes on to say:—"It is, I notice, a mistake to suppose that this plan of capturing the males can only be adopted in the breeding-season. The tame male can always be induced to 'buzz' by imitating the sound from some place hidden to him. This the Burmans do by twisting very rapidly between the palm of the hands a small stick, into a split at the top of which a piece of stiff cloth or a stiff leaf has been transversely inserted."

Nest.—A hollow scratched at the foot of a tree or in a clump of bamboo, more or less lined with dead leaves and a few feathers, and generally well-concealed.

Eggs.—Seven or eight in number, though as many as fifteen are said to be found at times; they vary in colour from pale cream to pinkish-buff; shell full of pores and without gloss. Average measurements, 1·97 by 1·46 inch.

SUB-SP. *a.* OATES' KALIJ PHEASANT. GENNÆUS OATESI.

Gennæus oatesi, Ogilvie-Grant, Cat. B. Brit. Mus. xxii. p. 306 (1893).

Adult Male.—Like the male of *G. lineatus*, but has the feathers of the lower back and rump *fringed with white;* only the margins of the inner webs of the middle pair of tail-feathers white without any black markings, while the white shaft-stripes on the sides of the breast are reduced in number or absent.

Adult Female.—Most like the female of *G. horsfieldi* in general plumage; but distinguished from this and the other allied species by having the outer tail-feathers *chestnut*, slightly mottled with black.

Range.—South-eastern Arakan Hills; in the vicinity of Prome and Thayetmyo.

VI. ANDERSON'S KALIJ PHEASANT. GENNÆUS ANDERSONI.

Euplocamus andersoni, Elliot, P. Z. S. 1871, p. 137; id. Monogr. Phasian. ii. pl. 22 (1872); Anderson, Res. Zool. Exped. Yun-nan, p. 670, pl. liii. (1878).

Euplocamus crawfurdi, Hume and Marshall (*nec* J. E. Gray), Game Birds of India, i. p. 203, pl. (1878).

Gennæus andersoni, Ogilvie-Grant, Cat. B. Brit. Mus. xxii. p. 306 (1893).

Adult Male.—Differs from the *male* of *G. lineatus* in having the feathers of the back and wing-coverts regularly marked *with about ten alternate, black and white, concentric bands*. Total length, 28 inches; wing, 9·9; tail, 13·6; tarsus, 3·4.

Adult Female.—Differs from the *female* of *G. lineatus* in having the white shaft-stripes on the feathers of the underparts *much* wider, and the outer webs of the secondary quills with irregular oblique buff bars, *reaching to the shaft*. Total length, 23·5 inches; wing, 9·3; tail, 9·7; tarsus, 3·1.

Range.—Kachin Hills east of Bhâmo and the Salween Valley as far south as Dargwin.

Very few examples of this fine Kalij have been procured, but it is not improbable that, when specimens (if such occur, as no doubt they do) are obtained from the intermediate parts of South China which lie between Yun-nan and Fo-kien, we may find that this form gradually grades into the Silver Kalij Pheasant.

VII. THE SILVER KALIJ PHEASANT. GENNÆUS NYCTHEMERUS.

White China Pheasant, Albin, Nat. Hist. B. iii. p. 35, pl. xxxvii. (1740).

Black and White Chinese Pheasant, Edwards, Nat. Hist. B. ii. pl. 66 (1747).

Phasianus nycthemerus, Linn. S. N. i. p. 272 (1766).

Euplocomus nycthemerus, J. E. Gray, Ill. Ind. Zool. ii. pl. 38, fig. 2 (1834); Gould, B. Asia, vii. pl. 17 (1859); Elliot, Monogr. Phasian. ii. pl. 21 (1872); David and Oustalet, Ois. Chine, p. 416 (1877).

Gennæus nycthemerus, Ogilvie-Grant, Cat. B. Brit. Mus. xxii. p. 307 (1893).

Adult Male.—Top of the head, long crest, and under-parts black, glossed with purple; upper-parts white, most of the feathers with five or six narrow, regular, black, concentric lines, fewer and less regular on the wing-coverts and quills; some of the feathers of the sides of the breast with white shaft-stripes, others with the whole of the outer webs white; tail longer than in the other species, the middle pair of feathers pure white, the outer pairs with oblique black lines. Total length, 40 inches; wing, 10·5; tail, 24; tarsus, 3·6.

Adult Female.—Crest blackish-brown; upper- and *under-parts* and middle pair of tail-feathers olive-brown, finely mottled with dusky lines; throat brownish-white; *outer tail-feathers black*, with *irregular oblique white lines*. Total length, 20·5 inches; wing, 9·1; tail, 9·8; tarsus, 3·2.

Range.—South China, Fo-kien and Che-kiang.

Habits.—According to Abbé David, the Silver Pheasant is becoming very rare in a wild state, and is only found in South China, towards the north of Fo-kien and perhaps in Che-kiang. He says that most of the Golden and Silver Pheasants that one sees at Shanghai come from Japan, where these two Chinese species are reared in captivity. The Silver Pheasant is known in China by the names of Ing-ky (Silver Fowl) and Paé-ky (White Fowl). Very little indeed is known of the habits of this extremely fine species in a wild state, though it has long been one of the commonest aviary birds. The males are unfortunately so extremely pugnacious and such big heavy birds that they fight with, and often kill, any other male Pheasant living in the same aviary, and for this reason must be kept separate.

Nest.—Like that of the other species.

Eggs.—Broad ovals; creamy-buff to brownish-buff, finely pitted all over and slightly glossed. Average measurements, 2·1 by 1·6 inches.

VIII. SWINHOE'S KALIJ PHEASANT. GENNÆUS SWINHOII.

Euplocamus swinhoii, Gould, P. Z. S. 1862, p. 284; id. B.
    Asia, vii. pl. 16 (1864); Sclater, in Wolf's Zool. Sketches
    (2), pl. 37 (1867); Elliot, Monogr. Phasian. ii. pl. 25 (1872);
    David and Oustalet, Ois. Chine, p. 417, pl. 102 (1877).
Gennæus swinhoii, Ogilvie-Grant, Cat. B. Brit. Mus. xxii. p.
    309 (1893).

Adult Male.—Feathered parts of the head, chin, and throat black; crest, mantle, and middle pair of tail-feathers *pure white*; scapulars *dark crimson* with bronzy-red reflections; rest of upper- and under-parts and outer tail-feathers black, glossed with purplish-blue, especially on the chest and breast. Total length, 29·5 inches; wing, 9·5; tail, 16; tarsus, 3·8.

Adult Female.—Crest rather short; head, back, and wing-coverts reddish-brown, the former with rufous-buff shaft-stripes to the feathers, the latter with the middle of each feather black and a triangular yellowish-buff spot near the tip; rest of upper-parts black, closely mottled with buff; inner webs of primary quills with *wide alternate bars of chestnut and black*; throat whitish-brown; chest and breast pale brown and marked like the back; rest of under-parts rufous-buff irregularly mottled with black; outer tail-feathers dark chestnut with some black mottling. Total length, 19·6 inches; wing, 9·1; tail, 7·9; tarsus, 3·1.

Range.—Mountain forests of Formosa.

This species was discovered by the late Mr. R. Swinhoe, for many years H.M. Consul in Formosa, who gives the following account:—

Habits.—"I was informed by my hunters that a second species of Pheasant, which was denominated by the Chinese colonists Wá-koë, was found in the interior mountains; that it was a true jungle bird, frequenting the wild hill-ranges of the aborigines, and rarely descending to the lower hills that border on the Chinese territory, and that in the evening and early morning the male was in the habit of showing himself on an exposed branch or roof of a savage's hut, uttering his crowing defiant note, while he strutted and threw up his tail like a rooster. I offered rewards and encouraged my men to do their utmost to procure me specimens of this bird, and I was so far successful that I managed to obtain a pair, but in my trip to the interior it was in vain that I sought to get a view of it in its native haunts, and to make acquaintance with it in a state of nature."

Later on he writes: "This bird is rare, and extremely difficult to procure, as the mountain travelling here is far from safe. My chief bird-hunter was nearly murdered and robbed of fifty pounds the other day while in search of Deer and this Pheasant."

This species is now frequently brought to this country alive, and the male is one of the handsomest of aviary Pheasants.

Eggs.—Oval, somewhat pointed at the smaller end; buff-cream colour, very minutely dotted with white. Measurements, 2·4 by 1·7 inches.

THE KOKLASS PHEASANTS. GENUS PUCRASIA.

Pucrasia, G. R. Gray, List Gen. B. 2nd ed. p. 79 (1841).

Type, *P. macrolopha* (Lesson).

Tail composed of *sixteen* feathers, long and wedge-shaped, the middle pair of feathers rather the longest and about twice as long as the outer pair. Upper tail-coverts very long, more than half the length of the tail.

First primary flight-feather considerably shorter than the second, which is about equal to the eighth; fourth somewhat the longest.

Sides of the head feathered; feathers of the body long and pointed.

Male with an elongate crest (short in the female); the feathers behind the ear-coverts greatly elongate, surpassing the crest in length, and the feet armed with a fairly strong pair of spurs.

The Koklass Pheasants may be conveniently divided into two groups:

A. Basal part of the outer tail-feathers mostly black or black and chestnut, *never grey* (species 1 to 4, pp. 281-285).

B. Basal part of the outer tail-feathers grey (species 5, 6, pp. 285, 286).

PLATE XXI.

COMMON KOKLASS PHEASANT.

A. Basal part of the outer tail-feathers mostly black or black and chestnut, never grey.

1. THE COMMON KOKLASS PHEASANT. PUCRASIA MACROLOPHA.

Satyra macrolopha, Less. Dict. Sci. Nat. lix. p. 196 (1829).
Phasianus pucrasia, J. E. Gray, Ill. Ind. Zool. i. pl. 40 (1830-32); Gould, Cent. B. Himal. pls. 69, 70 (1832).
Pucrasia macrolopha, G. R. Gray, List Gall. Brit. Mus. p. 31 (1844); Gould, B. Asia, vii. pl. 26 (1854); Elliot, Monogr. Phasian. i. pl. 28 (1872); Hume and Marshall, Game Birds of India, i. p. 159, pl. (1878); Ogilvie-Grant, Cat. B. Brit. Mus. xxii. p. 311 (1893).

(*Plate XXI.*)

Adult Male.—A long buff occipital crest; a large white patch on each side of the neck; rest of head and neck black, glossed with dark green, the feathers behind the ear-coverts being enormously elongate and longer than the crest; general colour of the upper-parts and sides grey, brownish on the wings, most of the feathers with black shaft-stripes; middle of the chest and under-parts dark chestnut; *outer tail-feathers black, shading into rufous on the basal half of the outer web*, and tipped with white. Total length, 23 inches; wing, 9·4; tail, 9·5; tarsus, 2·6.

Adult Female.—General colour above black, including a short crest, mottled with sandy-buff, most of the feathers with well-marked pale reddish-buff shaft-stripes; chin and throat whitish; under-parts pale rufous, edged and mottled on the breast and sides with black; outer tail-feathers *mostly black*, chestnut towards the base, and tipped with white. Total length, 19·6 inches; wing, 8·7; tail, 7·0; tarsus, 2·2.

Range.—Forests of the Western Himalayas from Kumaon to Chamba.

Habits.—Wilson says of the Koklass:

"This is another forest Pheasant common to the whole of the wooded regions, from an elevation of about 4,000 feet to nearly the extreme limits of forest, but is most abundant in the lower and intermediate ranges. In the lower ranges its favourite haunts are in wooded ravines; but it is found on nearly all hillsides which are covered with trees or bushes, from the summit of the ridges to about half way down. Farther in the interior it is found scattered in all parts, from near the foot of the hills to the top, or as far as the forest reaches, seeming most partial to the deep sloping forests composed of oak, chestnut, and morenda-pine, with box, yew, and other trees intermingled, and a thick underwood of ringal.

"The Koklass is of a rather retired and solitary disposition. It is generally found singly or in pairs; and, except the brood of young birds, which keep pretty well collected till near the end of winter, they seldom congregate much together. When numerous, several are often put up at no great distance from each other, as if they were members of one lot; but when more thinly scattered, it is seldom that more than two old birds are found together; and at whatever season, when one is found, its mate may, almost to a certainty, be found somewhere near. This would lead one to imagine that many pairs do not separate after the business of incubation is over, but keep paired for several successive years.

"In forests where there is little grass or underwood, they get up as soon as aware of the approach of anyone near, or run quickly along the ground to some distance; but where there is much cover they lie very close, and will not get up till forced by dogs or beaters. When put up by dogs they often fly up into a tree close by, which they rarely do when flushed by beaters or the sportsman himself, then flying a long way, and generally alighting on the ground. Their flight is rapid in the extreme, and after a few whirs, they sometimes shoot down like lightning. They now and then utter a few low chuckles before getting up, and occasionally rise with a low screeching chatter, and sometimes silently. The males often crow at daybreak, and occasionally at all hours.

"In the remote forests of the interior, on the report of a gun, all which are within half a mile or so, will often crow after each report. They also often crow after a clap of thunder or any loud and sudden noise; this peculiarity seems to be confined to those in dark shady woods in the interior, as I never noticed it on the lower hills.

"The Koklass feeds principally on leaves and buds; it also eats roots, grubs, acorns, seeds and berries, moss and flowers. It will not readily eat grain, and is more difficult to rear in confinement than the Jewar or Moonal. It roosts in trees generally, but at times on low bushes or on the ground.

"In the lower regions this bird should be sought for from about the middle of the hill upwards; oak forests, where the ground is rocky and uneven, are the most likely places to find it. Dogs are requisite to ensure sport, and are much to be preferred to beaters, as birds which, if flushed by the latter, would go far out of all reach, will often fly into the trees close above the dogs, and may be approached quite close, seeming to pay more attention to the movements of the dogs than to the presence of the sportsman. In the interior they will be found with the Moonal in all forests, but always keep in the wood, and do not, like it, resort to the borders. They are worth shooting, if but for the table, as the flesh is, perhaps, the best of the Hill Pheasants."

Captain Baldwin writes:—"The sportsman, on awakening in the early morning, when encamped on the uplands to hunt *Thar*, will hear the harsh '*Kokkok pokrass*' cry of this bird on all sides, and *Pucrasia macrolopha*, when heralding the dawn of day in this manner, is generally sitting on one of the lower boughs of a cypress-tree."

Nest.—A hole scraped in the ground, and sheltered by a tuft of grass or bush or rock, met with at elevations of from 5,000 to 11,000 feet.

Eggs.—Oval, somewhat pointed towards the small end; rich buff, finely or coarsely marked with brownish-red. They vary much in size: 1·85 to 2·29 by 1·39 to 1·57 inches. Average measurements, 2·08 by 1·47 inches.

SUB-SP. *a*. BIDDULPH'S KOKLASS PHEASANT. PUCRASIA BIDDULPHI.

Pucrasia biddulphi, Marshall, Ibis, 1879, p. 461; Ogilvie-Grant, Cat. B. Brit. Mus. xxii. p. 313 (1893).

Adult Male.—Differs from *P. macrolopha* in having the dark chestnut of the fore-neck extending more or less completely round the neck; the chestnut of the under-parts much darker and mixed with black.

Adult Female.—Like the *female* of *P. macrolopha*.

Range.—North-western Himalayas; Cashmere.

II. THE NEPAL KOKLASS PHEASANT. PUCRASIA NIPALENSIS.

Tragopan pucrasia, Temm. (*nec* J. E. Gray), Pl. Col. v. pl. 15 [No. 545] (1834).
Pucrasia nipalensis, Gould, P. Z. S. 1854, p. 100; id. B. Asia, vii. pl. 28 (1854); Hume and Marshall, Game Birds of India, i. p. 165, pl. (1878); Ogilvie-Grant, Cat. B. Brit. Mus. xxii. p. 314 (1893).
Pucrasia duvauceli, Elliot, Monogr. Phasian. i. pl. 29 (1872).

Adult Male.—Differs from *P. macrolopha* in having the feathers of the mantle, sides, and flanks *black, narrowly margined with grey;* black being the *predominating* colour of the upper-parts and sides of the body.

Adult Female.—Like the *female* of *P. macrolopha*, but rather richer in general colouring, especially on the under-parts.

Range.—Forests of the Central Himalayas; Western Nepal.

Practically nothing is known of the habits of this species and, so far as we are aware, it has never been shot by any European.

III. THE CHESTNUT-MANTLED KOKLASS PHEASANT. PUCRASIA
CASTANEA.

Pucrasia castanea, Gould, P. Z. S. 1854, p. 99; id. B. Asia,
vii. pl. 27 (1854); Ogilvie-Grant, Cat. B. Brit. Mus. xxii.
p. 314 (1893).

Adult Male.—Easily recognised from *P. macrolopha* by having
the *nape and upper-part of the mantle deep chestnut*, like the rest
of the under-parts.

Adult Female.—Has not yet been described.

Range.—Rather uncertain, but probably Northern Afghanistan and Kafiristan.

I have begged several of the officers about to take part in
the present disturbances at Chitral to look out for this fine
bird; whether they will have any opportunity of obtaining specimens remains to be seen, but I sincerely hope that they may.

IV. MEYER'S KOKLASS PHEASANT. PUCRASIA MEYERI.

Pucrasia meyeri, Madarász, Ibis, 1886, p. 145; Ogilvie-Grant,
Cat. B. Brit. Mus. xxii. p. 315 (1893).

Adult Male.—Distinguished from the male of *P. macrolopha*
by having *a well-marked yellow nuchal collar* like that of the
next species, *P. xanthospila*, from which it differs in the colour
of the outer tail-feathers, which are mostly rufous.

Adult Female.—Like the *female* of the next species, *P. xanthospila*, but the outer tail-feathers are mostly rufous.

Range.—Yer-ka-lo on the Upper Mékong to Central Tibet.

B. Basal part of the outer tail-feathers grey.

V. THE YELLOW-NECKED KOKLASS PHEASANT. PUCRASIA
XANTHOSPILA.

Pucrasia xanthospila, Gray, P. Z. S. 1864, p. 259, pl. xx.;
Milne-Edwards, N. Arch. Mus. Bull. i. p. 14, pl. 1, figs. 3
and 4 (1865); Gould, B. Asia, vii. pl. 24 (1869); Elliot,
Monogr. Phasian. i. pl. 30 (1872); David and Oustalet,
Ois. Chine, p. 407, pl. 104 (1877); Ogilvie-Grant, Cat. B.
Brit. Mus. xxii. p. 315 (1893).

Pucrasia xanthospila, var. *ruficollis*, David and Oustalet, Ois. Chine, p. 408 (1877).

Adult Male.—Like *P. macrolopha*, and especially *P. meyeri*, in general appearance, but the arrangement of the colours on the upper-parts, sides, and flanks is reversed, the *shaft-stripes* and *margins* of the feathers being grey and divided from one another by a *black band; a yellowish-buff or rufous-buff nuchal collar;* outer tail-feathers *grey*, with several black bands, the widest near the tip.

Adult Female.—Easily distinguished from the *female* of *P. macrolopha* by the grey outer tail-feathers, barred with black.

Range.—Mountain forests of North-western China, extending into Manchuria and Eastern Tibet.

Habits.—According to Abbé David the Yellow-necked Koklass, or *Song-ky* (Pine-Fowl), is found in small numbers in the wooded mountains of North-west China, and extends to Manchuria and Eastern Tibet. They never stray far from the underwood and jungle, where they are found solitary or in couples, feeding on grain and various vegetation, especially conifers. Their habits are like those of the True Pheasants, and they are excellent eating, being much superior to the other birds of the Pheasant-tribe met with in that part of China.

VI. DARWIN'S KOKLASS PHEASANT. PUCRASIA DARWINI.

Pucrasia darwini, Swinhoe, P. Z. S. 1872, p. 552; Elliot, Monogr. Phasian. i. pl. 30 *bis* (1872); Gould, B. Asia, vii. pl. 25 (1875); David and Oustalet, Ois. Chine, p. 409 (1877); Ogilvie-Grant, Cat. B. Brit. Mus. xxii. p. 316 (1893).

Adult Male.—Distinguished from the male of *P. xanthospila* by having no yellowish-buff or rufous-buff collar, and the *ground-colour of the sides and flanks pale reddish-buff*, though the disposition of the black marking on these parts and the mantle is perfectly similar.

Adult Female.—Much like the *female* of *P. xanthospila*, but the black bars on the outer tail-feathers are incomplete and represented by black spots on the shaft.

Range.—Mountain forest of Eastern China, Ngan-whi, Che-kiang, and Fo-kien.

Habits.—This very distinct Koklass is a resident in the above-mentioned mountains, and is pretty common in Fo-kien. Like the Yellow-necked Koklass, it is called *Song-ky* (Pine-Fowl) by the natives and is met with as a rule singly in the steep wooded mountains. Its food and habits are quite similar to those of its ally.

END OF VOL. I.

APPENDIX.

p. 128, add:—

XXXII*a*. DYBOWSKI'S FRANCOLIN.   FRANCOLINUS DYBOWSKII.

Francolinus dybowskii, Oust. Le Nat. (2), xiv. p. 232 (1893).

Adult.—Said by Dr. Oustalet to be allied to *F. gedgii* and *F. hartlaubi*, but to have buff bands on the outer web and on the margin of the inner web of the primaries; the forehead dusky, and the lores white.

Range.—Bangui, Upper Congo.

ALPHABETICAL INDEX.

acatoptricus, Tetrao. 48.
Acomus. 240.
  erythrophthalmus. 240.
  inornatus. 241, 242.
  pyronotus. 242.
adansoni, Excalfactoria. 197.
adansonii, Coturnix. 197.
  Perdix. 126.
Adanson's Painted Quail. 197.
adspersa, Scleroptera. 124.
adspersus, Francolinus. 124.
afer, Francolinus. 117.
  Pternistes. 137.
  Tetrao. 137.
African Painted Sand-Grouse. 24.
African Spur-Fowl. 78.
  Stone Pheasant. 199.
africanus, Francolinus 117.
Ahanta Francolin. 133.
ahantensis, Francolinus. 133.
albocristatus, Euplocamus. 258
  Gallophasis. 258, 262, 264, 271.
  Gennæus. 258.
  Phasianus. 258.
albogularis, Francolinus. 115.
albus, Lagopus. 36.
  Tetrao. 36.
alchata, Pterocles. 8.
  Pteroclurus. 8.
  Tetrao. 8.
alchatus, Pteroclurus. 2, 3, 8, 14.
Alectrophasis pyronota. 242.
alleni, Lagopus. 38.
alpina, Chourtka. 90.
  Lagopus. 43.
Altai Snow-Cock. 86.
altaicus, Perdix. 86.
  Tetraogallus. 86.
altumi, Francolinus. 131.
American Capercailzies. 58.
  Grouse. 59.
  Partridges. 78.

americana, Cupidonia. 64.
americanus Tympanuchus. 62, 65, 66.
Ammoperdix. 99.
  bonhami. 99, 101.
  heyi. 101.
andersoni, Euplocamus. 276
  Gennæus. 276.
Anderson's Kalij. 266, 276.
Aracan Tree-Partridge. 165.
Arboricola. 160, 161.
  ardens. 164.
  atrigularis. 163, 166.
  bambusæ. 203
  brunneipectus. 169, 173.
  charltoni. 174.
  chloropus. 172.
  crudigularis. 164.
  erythrophrys. 171.
  gingica. 166.
  hyperythra. 170, 171.
  intermedia. 165.
  javanica. 167.
  mandellii. 167.
  orientalis. 171.
  rubrirostris. 168.
  rufigularis. 165, 166, 173.
  sumatrana. 172.
  torqueola. 160, 165.
Arborophila mandellii. 167.
  sumatrana. 172.
ardens, Arboricola. 164.
arenarius, Pterocles. 11, 13, 15
  Tetrao. 15.
argoondah, Coturnix. 155.
  Perdicula. 153, 155.
argus, Argusianus. 199.
Argus Pheasants. 79, 199.
Argusianus. 79.
  argus. 199
atkhensis, Lagopus rupestris. 42.
auritum, Crossoptilon. 82, 248, 252, 254, 255, 257.

auritus, Phasianus. 255.
Australian Quail. 187, 190.
australis, Excalfactoria. 196.
  Synoicus. 190, 192.
ashantensis, Francolinus. 134.
asiatica, Perdicula. 153.
  Perdix. 153, 155.
atrigularis, Arboricola. 163, 166.

Bamboo Partridges. 78.
Bamboo-Pheasant. 202.
  Chinese. 203.
  Formosan. 204.
  Fytch's. 202.
bambusae, Arboricola. 203.
Bambusicola. 78, 202.
  fytchii. 202.
  hopkinsoni. 202.
  hyperythra. 170.
  sonorivox. 204.
  thoracica. 202, 203.
Barbary Red-legged Partridge. 97.
barbata, Perdix. 97, 149.
Bare-throated Francolins. 135.
Bearded Partridge. 149.
betulina, Bonasa. 75.
betulinus, Tetrao. 74.
biddulphi, Pucrasia. 284.
Biddulph's Koklass Pheasant. 284.
bicalcarata, Galloperdix. 210.
bicalcaratus, Francolinus. 126.
  Perdix. 210.
  Tetrao. 126.
bicinctus, Œnas. 24.
  Pterocles. 21.
Black and White Chinese Pheasant. 277.
Black-backed Kalij Pheasant. 263.
Black-bellied Sand-Grouse. 11, 15.
Black-breasted Kalij. 266, 269.
Black-breasted Quail. 185.
Black-cock. 36, 41.
Black Crestless Fire-Back. 242.
Black Grouse. 45, 52, 53.
  Caucasian. 46, 48.
Black-headed Red-legged Partridge. 98.
Black-throated Tree-Partridge. 163.
Black Wood-Partridge. 178, 179.
  witti, Microperdix. 158.
Blewitt's Painted Bush-Quail. 158.
Blood Pheasant. 212, 214, 215.
  Geoffroy's. 218.

Blood Pheasant, Northern. 219.
Blue Grouse. 59.
blythi, Ceriornis. 228.
  Tragopan. 228.
Blyth's Horned Pheasant. 228.
boehmi, Pternistes. 138.
Boehm's Bare-throated Francolin. 138.
Bonasa. 71.
  betulina. 75.
  sabinii. 72.
  umbelloides. 72.
  umbellus. 71.
  umbellus sabini. 71.
  umbellus togata. 71.
  umbellus umbelloides. 71.
Bonasia sylvestris. 74.
bonasia, Tetrao. 74.
  Tetrastes. 74, 75.
bonhami, Ammoperdix. 99, 101.
  Caccabis. 99.
  Perdix. 99.
Bonham's Seesee Partridge. 99.
Bornean Crestless Fire-Back. 242, 246.
Bornean Ferruginous Wood-Partridge. 176.
borneensis, Caloperdix. 176.
Bridled Sand-Grouse. 16.
Brown-breasted Tree-Partridge. 169.
brunneipectus, Arboricola. 169, 173.
buckleyi, Francolinus. 112.
bulweri, Lobiophasis. 248, 249, 250.
Bulwer's Wattled Pheasant. 249.
Bush-Quail, Blewitt's Painted. 158.
  Jungle. 153.
  Manipur Painted. 159.
  Painted. 156.
  Rock. 155.
Büttikofer's Francolin. 121.

Cabanis' Bare-throated Francolin. 141.
caboti, Ceriornis. 229.
  Tragopan. 229.
Cabot's Horned Pheasant. 229.
Caccabis. 90.
  bonhami. 99.
  chukar. 81, 91, 92, 95, 200.
  heyii. 101.

Caccabis magna. 95.
 melanocephala. 98.
 petrosa. 97.
 rufa. 95, 96, 98.
 rufa hispanica. 96.
 saxatilis. 90, 91, 92.
caineana, Coturnix. 193.
Caloperdix. 175.
 borneensis. 176.
 oculea. 175, 176, 177.
 sumatrana. 176.
cambayensis, Perdix. 153.
Canace canadensis. 54, 56, 58.
 franklini. 56.
 fuliginosus. 60.
 obscurus. 60.
Canachites. 54.
 canadensis. 54, 55.
 franklini. 56.
Canada Grouse. 54, 56, 58.
canadensis, Canace, 54, 56, 58.
 Canachites. 54, 55.
 Dendragapus. 55.
 Tetrao. 54.
Canadian Grouse. 54.
Cape Bare-throated Francolin. 136.
Cape Francolin. 129.
 Quail. 183.
capensis, Coturnix. 181, 183.
 Francolinus. 129, 139.
 Tetrao. 129.
Capercailzie. 48, 49, 53.
 American. 58.
 Dusky. 58, 59, 61.
 Kamtschatkan. 54.
 Richardson's. 61.
 Slender-billed. 53.
 Sooty. 60.
 Ural. 52.
Caspian Snow-Cock. 89.
caspius, Tetrao. 89, 90.
 Tetraogallus. 89.
castanea, Pucrasia. 285.
castaneicaudatus, Lobiophasis. 249.
castaneicollis, Francolinus. 118.
Caucasian Black Grouse. 46, 48.
Caucasian Snow-Cock. 90.
caucasica, Megaloperdix. 90.
 Tetrao. 90.
caucasicus, Tetraogallus. 90.
Centrocercus. 66.
 urophasianus. 66, 67.

Ceriornis blythi. 228.
 caboti. 229.
 melanocephala. 224.
 modestus. 229.
 satyra. 220.
 temmincki. 227.
cervinus, Synoicus. 190.
Ceylon Spur-Fowl. 210.
Chalcophasis sclateri. 240.
challayei, Tetraogallus. 89.
Chamba Moonal Pheasant. 237.
chambanus, Lophophorus. 232, 237, 238.
charltoni, Arboricola. 174.
 Perdix. 173.
 Tropicoperdix. 173, 174.
Charlton's Wood-Partridge. 173.
Chestnut-mantled Koklass Pheasant. 285.
Chestnut-naped Francolin. 118.
chinensis, Excalfactoria. 193, 194, 196, 198.
 Francolinus. 107, 205.
 Tetrao. 107, 193.
Chinese Bamboo-Pheasant. 203.
Chinese Francolin. 107, 205.
 Quail. 193.
chloropus, Arboricola. 172.
 Tropicoperdix. 172.
Chourtka alpina. 90.
chucar, Perdix. 91.
chukar, Caccabis. 81, 91, 92, 95, 200.
Chukar, Red-legged. 91.
Chukor. 81, 200.
cinerea, Perdix. 143, 149.
 Starna. 148, 149.
clamator, Francolinus. 129.
clappertoni, Francolinus. 126, 127, 128.
 Perdix. 126.
Clapperton's Francolin. 126.
Close-barred Francolin. 124.
 Sand-Grouse. 20.
Cock, Jungle. 270.
 Sage. 59.
 Snow, 83.
coerulescens, Crossoptilon. 255.
Columbæ. 1.
Columba phæonota. 20.
Columbian Sharp-tailed Grouse. 69.
columbianus, Pediœcetes. 69.

ALPHABETICAL INDEX.

columbianus, Phasianus. 69.
Common Koklass Pheasant. 281.
  Moonal Pheasant. 231.
  Partridge. 143, 147, 203.
  Pin-tailed Sand-Grouse. 12.
  Ptarmigan. 38.
  Red-legged Partridge. 96.
  Tree-Partridge. 160.
communis, Coturnix. 180.
coqui, Francolinus. 111, 112.
  Perdix. 111.
coromandelica, Coturnix. 185.
  Perdix. 185.
Corn-Crake. 5.
cornutus, Phasianus. 220.
coronatus, Pterocles. 15.
Coronetted Sand-Grouse. 18.
Coturnix. 79, 179.
  adansonii. 197.
  argoondah. 155.
  caineana. 193.
  capensis. 181, 183.
  communis. 180.
  coromandelica. 185.
  crucigera. 187.
  dactylisonans. 180.
  delegorguei. 187.
  emini. 197.
  erythrorhyncha. 156.
  flavipes. 193.
  fornasini. 187.
  histrionica. 187.
  japonica. 181, 184, 187.
  novæ-zealandiæ. 188.
  pectoralis. 187, 188.
  pentah. 153.
  raalteni. 192.
  textilis. 185.
  vulgaris. 180.
  vulgaris japonica. 184.
coturnix, Coturnix. 179, 180, 184, 187.
  Perdix. 180.
  Tetrao. 180.
cranchii, Perdix. 138.
  Pternistes. 138.
Cranch's Bare-throated Francolin. 138.
crawfurdi, Euplocamus. 270.
Crested Wood-Partridges. 177.
Crestless Fire-Backed Pheasants. 240, 243.
Crex crex. 5.

Crimson-headed Wood-Partridges. 174.
Crimson Horned Pheasant. 220.
cristatus, Phasianus. 177.
Crossoptilon. 251.
  auritum. 82, 248, 252, 254, 256, 257.
  cœrulescens. 255.
  drouynii. 252.
  harmani. 257.
  leucurum. 253.
  manchuricum. 254.
  mantchuricum. 254.
  · tibetanum. 251, 253.
crucigera, Coturnix. 187.
crudigularis, Arboricola. 164.
  Oreoperdix. 164.
cruentus, Ithagenes. 214, 215, 219.
  Phasianus. 215, 237.
Cryptonyx dussumieri. 179.
  ferrugineus. 179.
  niger. 179.
cupido, Cupidonia. 62.
  Tetrao. 62, 65.
  Tympanuchus. 61, 65.
Cupidonia americana. 62.
  cupido. 62, 65.
  pallidicinctus. 65.
curvirostris, Tetrao. 142.
  Phasianus. 237.
cuvieri, Gennæus. 266, 268, 269, 271.
  Lophophorus. 271.
Cuvier's Kalij Pheasant. 271.

dactylisonans, Coturnix. 180.
damascena, Perdix. 148.
darwini, Pucrasia. 287.
Darwin's Koklass Pheasant. 287.
daurica, Perdix. 149.
daurica, Tetrao. 149.
davisoni, Gennæus. 268, 271.
Davison's Kalij Pheasant. 271.
decoratus, Pterocles. 16.
delegorguei, Coturnix. 187.
Delegorgue's Quail. 187.
De l'Huy's Moonal Pheasant. 238.
Dendragapus. 45, 58.
  canadensis. 55.
  franklinii. 56.
  fuliginosus. 59, 60, 61.
  obscurus. 58, 61.
  obscurus fuliginosus. 60.

Dendragapus obscurus richardsonii. 61.
  richardsoni. 61.
desgodinsi, Tetraophasis. 83.
diardi, Euplocomus. 247.
  Lophura. 247.
Diardigallus fasciolatus. 247.
  prælatus. 247.
Diard's Crested Fire-Back. 247.
diemenensis, Synoicus. 190.
Double-banded Sand-Grouse. 21.
Double-spurred Francolin. 126.
drouynii, Crossoptilon. 252.
dulitensis, Rhizothera. 142.
Dusky Capercailzie. 58, 59, 61.
  Pheasant-Grouse. 81.
dussumieri, Cryptonyx. 179.

Eared Pheasant. 82, 248, 251.
  Harman's. 257.
  Hodgson's. 252.
  Manchurian. 254.
  Pallas'. 255.
  White-tailed. 253.
Eastern Pin-tailed Sand-Grouse. 218.
elgonensis, Francolinus. 122.
Elgon Francolin. 122.
emini, Coturnix. 197.
erckeli, Francolinus. 124, 135.
erckelii, Perdix. 135.
Erckel's Francolin. 135.
erythrophrys, Arboricola. 171.
erythrophthalmus, Acomus. 240, 241.
  Euplocamus. 241, 242.
  Phasianus. 241.
erythrorhyncha, Microperdix. 156, 158.
  Perdicula. 156.
erythrorhynchus, Ptilopachus. 200.
Euplocamus albocristatus. 258.
  andersoni. 276.
  crawfurdi. 276.
  diardi. 247.
  erythrophthalmus. 241, 242.
  horsfieldi. 269.
  ignitus. 244.
  leucomelanus. 262.
  lineatus. 272.
  melanonotus. 263.
  nobilis. 246.
  nycthemerus. 277.

Euplocamus sumatranus. 244.
  swinhoii. 278.
Euplocomus prælatus. 247.
  pyronotus. 242.
  vieilloti. 244.
Excalfactoria. 193, 248.
  adansonii. 197.
  australis. 196.
  chinensis. 193, 194, 196, 198.
  lineata. 194, 196, 198.
  minima. 193, 194.
exustus, Pterocles. 12.
  Pteroclurus. 1, 2, 12, 23.

Falcipennis. 57.
  hartlaubi. 57.
falcipennis, Falcipennis. 57.
  Tetrao. 57.
fasciata, Tringa. 22.
fasciatus, Phasianus. 272.
  Pterocles. 22.
fasciolatus, Diardigallus. 247.
ferrugineus, Cryptonyx. 179.
Ferruginous Wood-Partridges. 175.
finschi, Francolinus. 118.
Finsch's Francolin. 118.
Fire-Back, Black Crestless. 242.
  Bornean Crested. 246.
  Bornean Crestless. 242.
  Diard's Crested. 247.
  Malayan Crested. 244.
  Malayan Crestless. 241.
fischeri, Francolinus. 132.
Fischer's Francolin. 132.
flavipes, Coturnix. 193.
Formosan Bamboo-Pheasant. 204.
  Tree-Partridge. 164.
fornasini, Coturnix. 187.
Francolin. 101.
  Ahanta. 133.
  Bare-throated. 135.
  Boehm's Bare-throated. 138.
  Büttikofer's. 121.
  Cabanis' Bare-throated. 141.
  Cape. 129.
  Cape Bare-throated. 136.
  Chestnut-naped. 118.
  Chinese. 107, 205.
  Clapperton's. 126.
  Close-barred. 124.
  Coqui. 111.
  Cranch's Bare-throated. 138.
  Double-spurred. 126.

ALPHABETICAL INDEX.   293

Francolin, Dybowski's, 287.
  Elgon. 122.
  Erckel's. 135.
  Finsch's. 118.
  Fischer's. 132.
  Gariep. 120.
  Gedge's. 127.
  Grant's. 114.
  Gray's Bare-throated, 140.
  Grey. 108.
  Grey-striped. 125.
  Harris's. 115.
  Hartlaub's. 127.
  Heuglin's Double-spurred. 128.
  Hildebrandt's. 131.
  Hubbard's. 112.
  Humboldt's Bare-throated. 136.
  Indian Swamp. 122, 123.
  Jackson's. 134.
  Johnston's. 132.
  Kirk's. 114.
  Latham's. 108.
  Levaillant's. 119.
  Long-billed. 141, 142.
  Natal. 130.
  Painted. 106.
  Pearl-breasted. 117.
  Reichenow's Bare-throated. 140.
  Ring-necked. 112.
  Rüppell's. 116.
  Scaled. 132.
  Schlegel's. 112.
  Schuett's. 133.
  Sclater's Bare-throated. 137.
  Sharpe's. 116, 128.
  Shelley's. 121.
  Smith's. 113.
  Spotted. 114.
  Swainson's Bare-throated. 139.
  Ulu. 117.
  White-throated. 115.
Francolinus. 101, 135.
  adspersus. 124.
  afer. 117.
  africanus. 117.
  ahantensis. 133.
  albigularis. 115.
  albogularis. 115.
  altumi. 131.
  ashantensis. 134.
  bicalcaratus. 126.
  buckleyi. 112.
  capensis. 129, 139.

Francolinus castaneicollis. 118.
  chinensis. 107, 205.
  clamator. 129.
  clappertoni. 126, 127, 128.
  coqui. 111, 112.
  dybowskii. 287.
  elgonensis. 122.
  erckeli. 124, 135.
  francolinus. 101, 106.
  finschi. 118.
  fischeri. 132.
  gariepensis. 120, 121.
  gedgii. 127.
  granti. 114, 115.
  griseostriatus. 125.
  gularis. 122, 123.
  gutturalis. 116.
  hartlaubi. 127, 128.
  hildebrandti. 132.
  hubbardi. 112.
  humboldti. 136.
  jacksoni. 134.
  johnstoni. 132.
  jugularis. 121.
  kirki. 114, 115.
  lathami. 108.
  leucoparæus. 137.
  leucoscepus. 140.
  levaillanti. 119.
  longirostris. 142.
  maculatus. 107.
  modestus. 133.
  natalensis. 130.
  nivosus. 209.
  nudicollis. 136.
  ochrogaster. 114.
  peli. 108.
  petiti. 133.
  phayrei. 107.
  pictus. 106, 107.
  pileatus. 113.
  pondicerianus. 108.
  psilolæmus. 115.
  rueppellii. 128.
  schlegelii. 112.
  schoanus. 114.
  schuetti. 133.
  sephæna. 113, 114, 115.
  sharpii. 116, 128.
  shelleyi. 121, 122.
  spadiceus. 206.
  spilogaster. 114.
  spilolæmus. 115, 116.

Francolinus squamatus. 132, 133.
  streptophorus. 112, 113.
  stuhlmanni. 111.
  subtorquatus. 111.
  swainsoni. 139.
  uluensis. 117.
francolinus, Francolinus. 101, 106.
franklini, Canace. 56.
  Canachites. 56.
  Tetrao. 56.
franklinii, Dendragapus. 56.
Franklin's Grouse. 56.
fuliginosus, Canace. 60.
  Dendragapus. 59, 60, 61.
fusca, Perdix. 199.
fuscus, Ptilopachus. 199, 206.
fytchii, Bambusicola. 202.
Fytch's Bamboo-Pheasant. 202.

Gallinæ. 1, 25.
Galloperdix. 78, 205.
  bicalcarata. 210.
  lunulata. 206.
  lunulatus. 208, 209.
  lunulosa. 209.
  spadicea. 206.
  spadiceus. 206.
Gallophasis albocristatus. 258, 262, 264, 271.
  horsfieldi. 269, 271.
  leucomelanus. 262.
  melanotus. 263.
Gallus gallus. 270.
  macartneyi. 246.
gallus, Gallus. 270.
Game-Birds. 1, 47.
gardneri, Phasianus. 215.
Gariep Francolin. 120.
gariepensis, Francolinus. 120, 121
Gedge's Francolin. 127.
gedgii, Francolinus. 127.
Gennæus. 258.
  albocristatus. 258.
  andersoni. 276.
  cuvieri. 266, 268, 269, 271.
  davisoni. 268, 271.
  horsfieldi. 266, 269.
  horsfieldi lineatus. 269.
  leucomelanus. 262, 265.
  lineatus. 248, 266, 272, 276, 277.
  melanonotus. 263.
  muthura. 264.

Gennæus nycthemerus. 258, 277.
  oatesi. 266, 268, 269, 276.
  swinhoii. 278.
geoffroyi, Ithagenes. 218.
  Ithaginis. 219.
Geoffroy's Blood Pheasant. 218.
gingica, Arboricola. 166.
gingicus, Tetrao. 166.
granti, Francolinus. 114, 115.
Grant's Francolin. 114.
Gray's Bare-throated Francolin. 140.
Grey-bellied Hazel-Hen. 77.
  Tragopan. 228.
Grey Francolin. 108.
Grey-striped Francolin. 125.
Grey Swamp-Quail. 192.
griseiventris, Tetrastes. 77.
griseogularis, Perdix. 99.
griseostriatus, Francolinus. 125.
Grouse. 26.
  American. 59.
  Black. 45, 47, 52, 53.
  Blue. 59.
  Canada. 54, 56, 58.
  Canadian. 54.
  Columbian Sharp-tailed. 69.
  Franklin's. 56.
  Hazel. 75, 79.
  Pheasant. 81.
  Pine. 59.
  Pinnated. 61.
  Red. 25, 27, 29, 42, 47.
  Ruffed. 71, 73.
  Sage. 66.
  Sand. 1, 3, 26.
  Sharp-tailed. 68.
  Sharp-winged. 57.
  Willow. 26, 27, 29, 36, 44, 52, 85.
Guinea-Fowls. 199.
gularis, Francolinus. 122.
  Ortygornis. 122.
  Perdix. 122.
gutturalis, Francolinus. 116.
  Perdix. 116.
  Pterocles. 19, 22.

Hæmatortyx. 174.
  sanguiniceps. 174.
Hainan Tree-Partridge. 164.
hamiltoni, Phasianus. 258.
hardwickii, Perdix. 208.

harmani, Crossoptilon. 257.
Harman's Eared-Pheasant. 257.
Harris's Francolin. 115.
hartlaubi, Falcipennis. 57.
  Francolinus. 127, 128.
Hartlaub's Francolin. 127.
hastingsi, Tragopan. 224.
Hawk-Eagle, Nepal. 225.
Hazel-Grouse. 75, 79.
Hazel-Hen. 48, 74.
  Grey-Bellied. 77.
  Severtzov's. 77.
Heath Hen. 65.
hemileucurus, Lagopus. 43.
Hemipodes. 26, 43.
Hen, Prairie. 59, 62.
henrici, Tetraogallus. 85.
hepburnii, Perdix. 106.
Heuglin's Double-spurred Francolin. 128.
heyi, Ammoperdix. 101.
  Caccabis. 101.
  Perdix. 101.
Hey's Seesee Partridge. 101.
hildebrandti, Francolinus. 132.
Hildebrandt's Francolin. 131.
Himalayan Snow-Cock. 86.
himalayensis, Tetraogallus. 83, 86, 89.
histrionica, Coturnix. 187.
hodgsoniæ, Perdix. 150, 151.
  Sacfa. 150.
Hodgson's Eared-Pheasant. 252.
hopkinsoni, Bambusicola. 202.
Horned Indian Pheasant. 220.
Horned Pheasant. 220.
  Blyth's. 228.
  Cabot's. 229.
  Temminck's. 227.
horsfieldi, Euplocamus. 269.
  Gallophasis. 269, 271.
  Gennæus. 266, 269.
Horsfield's Tree-Partridge. 171.
Hose's Long-billed Francolin. 142.
hubbardi, Francolinus. 112.
Hubbard's Francolin. 112.
humboldti, Francolinus. 136.
  Pternistes. 136.
Humboldt's Bare-throated Francolin. 136.
hybridus, Tetrao. 52.
hyperborea, Lagopus. 43.
hyperboreus, Lagopus. 43.

hyperythra, Arboricola. 170, 171.
  Bambusicola. 170.

icterorhynchus, Francolinus. 128.
ignita, Lophura. 246, 247.
ignitus, Euplocomus. 244.
  Phasianus. 244, 246.
impejanus, Phasianus. 237.
Impeyan Moonal Pheasant. 237.
impeyanus, Lophophorus. 231, 232, 236, 237.
Indian Bush-Quails. 153.
Indian Painted Sand-Grouse. 22.
Indian Spur-Fowl. 78.
Indian Swamp Francolin. 122.
infuscatus, Pternistes. 141.
inornatus, Acomus. 241, 242.
intermedia, Arboricola. 165.
Island Painted Quail. 196.
Ithagenes. 214.
  cruentus. 214, 215, 219.
  geoffroyi. 218, 219.
Ithaginis madagascariensis. 206.
  sinensis. 219.

jacksoni, Francolinus. 134.
Jackson's Francolin. 134.
Japanese Quail. 184.
japonica, Coturnix. 181, 184, 187.
javanica, Arboricola. 167.
javanicus, Tetrao. 167.
Javan Partridge. 167.
Javan Tree-Partridge. 167.
johnstoni, Francolinus. 132.
Johnston's Francolin. 132.
jugularis, Francolinus. 121.
Jungle Bush-Quail. 153.
Jungle Cock. 269.

Kalij Pheasant. 258.
  Anderson's. 266, 276.
  Black-backed. 263.
  Black-breasted. 266, 269.
  Cuvier's. 271.
  Davison's. 271.
  Oates'. 276.
  Silver. 277.
  Vermicellated. 266, 272.
Kamtschatkan Capercailzie. 54.
kamtschaticus, Tetrao. 54.
kennicotti, Pediocætes. 68, 69.
kirki, Francolinus. 114, 115.
Kirk's Francolin. 114.

Koklass Pheasant. 81, 280, 282.
   Chestnut-mantled. 285.

Lagopus. 26, 34, 45.
   albus. 36.
   alleni. 38.
   alpina. 43.
   hemileucurus. 43.
   hyperborea. 43.
   hyperboreus. 43.
   lagopus. 26, 36.
   leucurus. 44.
   mutus. 36, 38, 44.
   scoticus. 27, 36, 38.
   rupestris. 39, 42, 44.
   rupestris atkhensis. 42.
   rupestris nelsoni. 42.
   rupestris reinhardti. 42.
   welchi. 42.
lagopus, Lagopus. 26, 36.
   Tetrao, 36, 38.
La Perdrix rouge de Madagascar. 206.
La Petite Caille de l'Isle de Luçon. 196.
lathami, Francolinus. 108.
   Satyra. 220.
Latham's Chittygong Pheasant. 263.
   Francolin. 108.
Le Perdrix de Gingi. 166.
Lerwa. 79.
   lerwa. 79, 80.
   nivicola. 80.
lerwa, Lerwa. 79, 80.
Lesser Prairie Hen. 65.
leucomelanos, Phasianus. 262.
leucomelanus, Euplocamus. 262.
   Gallophasis. 262.
   Gennæus. 262, 265.
leucoparæus, Francolinus. 137.
   Pternistes. 137.
leucoscepus, Francolinus. 140.
   Pternistes. 140.
leucurum, Crossoptilon. 253.
leucurus, Lagopus. 44.
   Tetrao. 44.
levaillanti, Francolinus. 119.
levaillantii, Perdix. 119, 120.
Levaillant's Francolin. 119.
l'huysii, Lophophorus. 238.
lichtensteini, Pterocles. 20, 22.
lineata, Excalfactoria. 194, 196, 198.

Lineated Kalij Pheasant. 248.
lineatus, Euplocamus. 272.
   Gennæus. 248, 266, 269, 272, 276, 277.
   horsfieldi, Gennæus. 269.
   Oriolus. 196.
   Phasianus. 272.
Lobiophasis bulweri. 248, 250.
   castaneicaudatus. 249, 250.
Long-billed Francolins. 141, 142.
longirostris, Francolinus. 142.
   Perdix. 142.
   Rhizothera. 141.
Lophophorus. 230, 238.
   chambanus. 232, 237, 238.
   cuvieri. 271.
   impeyanus. 231, 232 236, 237.
   l'huysii. 238.
   mantoui. 236.
   obscura. 236.
   obscurus. 81, 236.
   sclateri. 230, 240.
   refulgens. 230, 231, 236.
Lophura diardi. 247.
   ignita. 246, 247.
   rufa. 243, 244, 247.
lucani, Pternistes. 138.
lunulata, Galloperdix. 206.
   Perdix. 208.
lunulatus, Galloperdix. 208, 209.
lunulosa, Galloperdix. 209.
Lyrurus. 45.
   mlokosiewiczi. 46, 48.
   tetrix. 36, 45, 49, 50, 56.

macartneyi, Gallus. 246.
macrolopha, Pucrasia. 280, 281, 283, 284.
   Satyra. 281.
maculatus, Francolinus. 107.
   Perdix. 107.
Madagascar Partridges. 151, 152.
madagascariensis, Ithaginis. 206.
   Margaroperdix. 151, 152.
   Tetrao. 107, 152.
magna, Caccabis. 95.
Malacortyx superciliaris. 213.
Malacoturnix superciliosus. 213.
Malayan Crested Fire-Back. 244.
   Crestless Fire-Back. 241.
Manchurian Eared-Pheasant. 254.
mandellii, Arboricola. 167.
   Arborophila. 167.

Mandelli's Tree-Partridge. 167.
manillensis, Tetrao. 196.
manipurensis, Microperdix. 159.
  Perdicula. 159.
Manipur Painted Bush-Quail. 159.
mantchuricum, Crossoptilon. 254.
mantoui, Lophophorus. 236.
Margaroperdix. 151.
  madagascariensis. 151, 152.
  striatus. 152.
Masked Sand-Grouse. 20.
medius, Tetrao. 48.
Megaloperdix caucasica. 90.
  raddei. 89.
megapodia, Perdix. 161.
melanocephala, Caccabis. 98.
  Ceriornis. 224.
  Perdix. 98.
  Satyra. 224.
melanocephalus, Phasianus. 224.
  Tragopan. 224.
melanonotus, Euplocomus. 263.
  Gennæus. 263.
Melanoperdix. 178.
  nigra. 179.
melanotus, Gallophasis. 263.
Meleagris satyra. 220.
meyeri, Pucrasia. 285.
Meyer's Koklass Pheasant. 285.
Microperdix. 156.
  blewitti. 158.
  erythrorhyncha. 156, 158.
  manipurensis. 159.
Migratory Partridge. 149.
  Quail. 180.
minima, Excalfactoria. 193, 194.
mlokosiewiczi, Lyrurus. 46, 48.
  Tetrao. 48.
modestus, Ceriornis. 229.
  Francolinus. 133.
montana, Perdix. 147.
Moonal Pheasant. 230.
  Chamba. 237.
  De l'Huy's. 238.
  Impeyan. 237.
  Sclater's. 240.
Mountain Pheasant-Quail. 213.
Mrs. Hodgson's Partridge. 150
muthura, Gennæus. 264.
  Phasianus. 263, 264.
mutus, Lagopus. 36, 38, 44.
  Tetrao. 38.

Namaqua Pin-tailed Sand-Grouse. 11.
namaqua, Pterocles. 11.
  Pteroclurus. 11.
  Tetrao. 11.
namaquus, Pteroclurus. 11.
Natal Francolin. 130.
natalensis, Francolinus. 130.
nelsoni, Lagopus rupestris. 42.
Nepal Hawk-Eagle. 225.
Nepal Kalij Pheasant. 292.
Nepal Koklass Pheasant. 284.
nepaulensis, Satyra. 220.
New Britain Painted Quail. 197.
New Zealand Quail. 188.
niger, Cryptonyx. 179.
nigra, Melanoperdix. 179.
nipalensis, Pucrasia. 284.
nivicola, Lerwa. 80.
nivosus, Francolinus. 209.
nobilis, Euplocomus. 246
Northern Blood Pheasant. 219.
northiæ, Polyplectron. 206.
novæ-zealandiæ, Coturnix. 188.
nudicollis, Francolinus. 136.
  Pternistes. 135, 136, 137.
  Tetrao. 136.
nycthemerus, Euplocomus. 277.
  Gennæus. 258, 277.
  Phasianus. 277.

oatesi, Gennæus. 266, 268, 269. 276.
Oates' Kalij Pheasant. 276.
obscura, Lophophorus. 236.
obscurus, Canace. 60.
  Dendragapus. 58, 61.
  fuliginosus, Dendragapus. 60.
  Lophophorus. 81, 236.
  richardsonii, Dendragapus. 61.
  Tetrao. 58.
  Tetraophasis. 81, 82.
ochrogaster, Francolinus. 114.
oculea, Caloperdix. 175, 176, 177.
  Perdix. 175.
Odontophorinæ. 78.
Œnas bicinctus. 24.
olivacea, Perdix. 161.
Ophrysia. 212.
  superciliosa. 212, 213
Oreoperdix erudigularis. 164.
orientalis, Arboricola. 171.
  Perdix. 169, 171.

Oriolus lineatus. 196.
Ortygornis gularis. 122.
   pondicerianus. 109.

Painted Bush-Quails. 156.
Painted Francolin. 106.
Painted Quails. 193, 248.
Painted Spur-Fowl. 208.
Pallas' Eared-Pheasant. 255.
   Sand-Grouse. 2.
   Three-toed Sand-Grouse. 3.
pallidicinctus, Cupidonia. 65.
paradoxa, Tetrao. 3.
paradoxus, Syrrhaptes. 2, 3, 4, 6, 7, 9.
Partridge. 73, 78.
   American. 78.
   Bamboo. 78.
   Bearded. 149.
   Bonham's Seesee. 99.
   Common. 143, 147, 203.
   Javan. 167.
   Madagascar. 151, 152.
   Migratory. 148.
   Mrs. Hodgson's. 150.
   Prjevalsky's. 151.
   Red-legged. 90.
   Rock Red-legged. 90.
   Seesee. 99.
   Snow. 79, 80.
   Tree. 160.
   True. 142.
   Wood. 172.
parvirostris, Tetrao. 53, 54.
Pearl-breasted Francolin. 117.
pectoralei, Coturnix. 187, 188.
Pediocætes kennicotti. 68, 69.
Pediocætes columbianus. 69.
   phasianellus. 68, 69, 70.
   phasianellus campestris. 69.
peli, Francolinus. 108.
Peloperdix rubrirostris. 168.
pennanti, Satyra. 220.
pentah, Coturnix. 153.
Perdicinæ. 78.
Perdicula. 153.
   argoondah. 153, 155.
   asiatica. 153.
   erythrorhyncha. 156.
   manipurensis. 159.
   rubicola. 153.
Perdix. 143.
   adansonii. 126.

Perdix altaicus. 86.
   asiatica. 153, 155.
   barbata. 97, 149.
   bicalcaratus. 210.
   bonhami. 99.
   cambayensis. 153.
   charltoni. 173.
   chucar. 91.
   cinerea. 143, 149.
   clappertoni. 126.
   coqui. 111.
   coromandelica. 185.
   coturnix. 180.
   cranchii. 138.
   damascena. 148.
   daurica. 149.
   erckelii. 135.
   fusca. 199.
   griseogularis. 99.
   gularis. 122.
   gutturalis. 116.
   hardwickii. 208.
   hepburnii. 106.
   heyi. 101.
   hodgsoniæ. 150, 151.
   levaillantii. 119, 120.
   longirostris. 142.
   lunulata. 208.
   maculatus. 107.
   megapodia. 161.
   melanocephala. 98.
   montana. 147.
   oculea. 175.
   olivacea. 161.
   orientalis. 109, 171.
   perdix. 143, 203.
   personata. 171.
   petrosa. 97.
   picta. 106.
   punctulata. 138.
   raaltenii. 192.
   robusta. 143.
   rubra. 96.
   rubricollis. 140.
   rupestris daurica. 149.
   saxatilis. 90.
   senegalensis. 126.
   sephæna. 113.
   sifanica. 151.
   sphenura. 203.
   striatus. 152.
   swainsoni. 139.
   thoracica. 203.

Perdix torqueola. 160.
  vaillanti. 119.
  ventralis. 199.
  zeylonensis. 210.
perdix, Perdix. 143, 203.
  Tetrao. 143, 149.
peregrina, Starna. 148.
perlatus, Tetrao. 107.
personata, Perdix. 171.
personatus, Pterocles. 20.
petiti, Francolinus. 133.
petrosa, Caccabis. 97.
  Perdix. 97.
petrosus, Tetrao. 97.
phæonota, Columba. 20.
phasianellus, Pediœcetes. 68, 69, 70.
  Tetrao. 68.
phasianellus campestris, Pediœcetes. 69.
Phasianidæ. 78, 199.
Phasianus. 78, 199.
  albocristatus. 258.
  auritus. 255.
  columbianus. 69.
  cornutus. 220.
  cristatus. 177.
  cruentus. 215.
  curvirostris. 237.
  erythrophthalmus. 241.
  fasciatus. 272.
  gardneri. 215.
  hamiltoni. 258.
  ignitus. 244, 246.
  impejanus. 237.
  leucomelanos. 262.
  lineatus. 272.
  melanocephalus. 224.
  muthura. 263.
  nycthemerus. 277.
  pucrasia. 281.
  purpureus. 241.
  reynaudii. 272.
  roulroul. 177.
  rufus. 244.
  tibetanus. 252.
phayrei, Francolinus. 107.
Pheasant. 48, 52, 78, 199.
  African Stone. 199.
  Argus. 79, 199.
  Bamboo. 202.
  Biddulph's Koklass. 284.
  Black and White Chinese. 277.

Pheasant, Blood. 212.
  Common Koklass. 281.
  Common Moonal. 231.
  Crestless Fire-backed. 240, 243
  Crimson Horned. 220.
  Darwin's Koklass. 287.
  Eared. 82, 248, 251.
  Horned. 220.
  Horned Indian. 220.
  Kalij. 258.
  Koklass. 280.
  Latham's Chittygong. 263.
  Lineated Kalij. 248.
  Meyer's Koklass. 285.
  Moonal. 230.
  Nepal Kalij. 262.
  Nepal Koklass. 284.
  Stone. 199.
  Swinhoe's Kalij. 278.
  Wattled. 248.
  Western Horned. 224.
  White China. 277.
  White-crested Kalij. 258.
  Yellow-necked Koklass. 286.
Pheasant-Grouse. 81.
  Dusky. 81.
  Széchenyi's. 83.
Pheasant-Quail. 212.
  Mountain. 213.
picta, Perdix. 106.
pictus, Francolinus. 106, 107.
pileatus, Francolinus. 113.
Pine Grouse. 59.
Pine-Hen. 59.
Pinnated Grouse. 61.
pintadeanus, Tetrao. 107.
Pin-tailed Four-toed Sand-Grouse. 7.
Pin-tailed Sand-Grouse. 2, 11.
  Common. 12.
plumbeus, Synœcus. 192.
Polyplectron northiæ. 206.
pondicerianus, Francolinus. 108.
  Ortygornis. 109.
  Tetrao. 108.
porphyrio, Tetrao. 177.
prælatus, Diardigallus. 247.
  Euplocomus. 247.
Prairie Hen. 62.
Prince Henry's Snow-Cock. 85.
Prjevalsky's Red-legged Partridge. 95.
psilolæmus, Francolinus. 115.

Ptarmigan. 26, 27, 29, 36, 39, 44.
  Common. 38.
  Rock. 42.
  Spitsbergen. 43.
  White-tailed. 44.
Pternistes. 135.
  afer. 137.
  boehmi. 138.
  cranchii. 138.
  humboldti. 136.
  infuscatus. 141.
  leucoparæus. 137.
  leucoscepus. 140.
  lucani. 138.
  nudicollis. 135, 136, 137.
  rubricollis. 137, 140.
  rufopictus. 140.
  sclateri. 137.
  swainsoni. 139.
Pterocles. 15.
  alchata. 8.
  arenarius. 11, 13, 15.
  bicinctus. 21.
  coronatus. 15, 18.
  decoratus. 16.
  exustus. 12.
  fasciatus. 22.
  gutturalis. 19, 22.
  lichtensteini. 20, 22.
  namaqua. 11.
  personatus. 20.
  pyrenaicus. 10.
  quadricinctus. 24.
  senegalus. 14.
  tricinctus. 24.
  variegatus. 17.
Pterocletes. 1.
Pteroclidæ. 1, 3.
Pteroclurus. 7.
  alchatus. 2, 3, 8, 14.
  exustus. 1, 2, 12, 23.
  namaqua. 11.
  namaquus. 11.
  pyrenaicus. 8, 10, 11.
  senegallus. 14, 18.
Ptilopachus. 199.
  erythrorhynchus. 200.
  fuscus. 199.
Ptilopachys. 78.
  fuscus. 200.
Pucrasia. 280.
  biddulphi. 284.
  castanea. 285.

Pucrasia darwini. 286.
  macrolopha. 81, 280, 281, 283, 284.
  meyeri. 285.
  nipalensis. 284.
  ruficollis. 286.
  xanthospila. 285, 286.
pucrasia, Phasianus. 281.
  Tragopan. 284.
punctulata, Perdix 138.
purpureus, Phasianus. 241.
pyrenaicus, Pterocles. 10.
  Pteroclurus. 8, 10, 11.
pyronota, Alectrophasis. 242.
pyronotus, Acomus. 242.
  Euplocomus. 242.

quadricinctus, Pterocles. 24.
Quail. 78, 79, 179.
  Adanson's Painted. 197.
  Australian. 187, 190.
  Black-breasted. 185.
  Cape. 183.
  Chinese. 193.
  Delegorgue's. 187.
  Indian Bush. 153.
  Island Painted. 196.
  Japanese. 184.
  Migratory. 180.
  New Britain Painted. 197.
  New Zealand. 188.
  Painted. 193, 248.
  Pheasant. 212.
  Rain. 185.
  Swamp. 190.

raalteni, Coturnix. 192.
  Synœcus. 191, 192.
raaltenii, Perdix. 192.
Raalten's Swamp-Quail. 192.
raddei, Megaloperdix. 89.
Rain-Quail. 185.
Red-billed Tree-Partridge. 168.
Red-crested Wood-Partridge. 177, 216.
Red Grouse. 25, 27, 29, 36, 42, 47.
Red-legged Partridge. 90.
  Barbary. 97.
  Black-headed. 98.
  Chukar. 91.
  Common. 95.
  Prjevalsky's. 95.

ALPHABETICAL INDEX.

Red Spur-Fowl. 206.
refulgens, Lophophorus. 230, 231, 236.
Reichenow's Bare-throated Francolin. 140.
reinhardti, Lagopus rupestris. 42.
reynaudii, Phasianus. 272.
Rhizothera. 141.
  dulitensis. 142.
  longirostris. 141.
richardsoni, Dendragapus. 61.
richardsonii, Tetrao. 61.
Richardson's Capercailzie. 61.
Ring-necked Francolin. 112.
Ripa. 36.
robusta, Perdix. 143.
Rock Bush-Quail. 155.
Rock Ptarmigan. 42.
Rock Red-legged Partridge. 90.
Rollulus. 177.
  roulroul. 177, 216.
  superciliosus. 213.
roulroul, Phasianus. 177.
  Rollulus. 177.
rubicola, Perdicula. 153.
rubra, Perdix. 96.
rubricollis, Perdix. 140.
  Pternistes. 137, 140.
  Tetrao. 137.
rubrirostris, Arboricola. 168.
  Peloperdix. 168.
rueppellii, Francolinus. 128.
rufa, Caccabis. 95, 98.
  Lophura. 243, 244, 247.
rufa hispanica, Caccabis. 96.
Ruffed Grouse. 71, 73.
ruficollis, Pucrasia. 286.
rufigularis, Arboricola. 165, 166, 173.
rufopictus, Pternistes. 140.
Rufous-throated Tree-Partridge. 165.
rufus, Phasianus. 244.
  Tetrao. 96.
rupestris, Lagopus. 39, 42, 44.
  Tetrao. 42.
rupestris atkhensis, Lagopus. 42.
rupestris daaurica, Perdix. 149.
  Starna. 149.
rupestris nelsoni, Lagopus. 42.
rupestris reinhardti, Lagopus. 42.
Rüppell's Francolin. 116.

sabinii, Bonasa. 72.
  Tetrao. 71.
Sacfa hodgsoniæ. 150
Sage Cock. 59.
Sage Grouse. 66.
saliceti, Tetrao. 36.
Sand-Grouse. 1, 3, 26.
  African Painted. 24.
  Black-bellied. 11, 15.
  Bridled. 16.
  Close-barred. 20.
  Coronetted. 18.
  Double-banded. 21
  Eastern Pin-tailed. 2, 8.
  Indian Painted. 22.
  Masked. 20.
  Namaqua Pin-tailed. 11.
  Pallas'. 2.
  Pallas' Three-toed. 3.
  Pin-tailed. 2.
  Pin-tailed Four-toed. 7.
  Short-tailed. 15.
  Smith's Chestnut-vented. 19.
  Spotted. 18.
  Spotted Pin-tailed. 14.
  Three-toed. 3.
  Variegated. 17.
  Western Pin-tailed. 8, 10.
sanguiniceps, Hæmatortyx. 174.
Satyra lathami. 220.
  macrolopha. 281.
  melanocephala. 224.
  nepaulensis. 220.
  pennanti. 220.
  temmincki. 227.
satyra, Ceriornis. 220.
  Meleagris. 220.
  Tragopan. 220, 224, 225, 227.
satyrus, Tragopan. 220.
saxatilis, Caccabis. 90, 91, 92.
  Perdix, 90.
Scaled Francolin. 132.
schlegeli, Francolinus. 112.
Schlegel's Francolin. 112.
schoanus, Francolinus. 114.
schuetti, Francolinus. 133.
Schuett's Francolin. 133.
sclateri, Chalcophasis. 240.
  Lophophorus. 230, 240.
  Pternistes. 137.
Sclater's Bare-throated Francolin. 137.
Sclater's Moonal Pheasant. 240.

Scleroptera adspersa. 124.
scoticus, Lagopus. 27, 36, 38.
   Tetrao. 27.
Seesee Partridges. 99.
   Hey's. 101.
senegalensis, Perdix. 126.
senegallus, Pteroclurus. 14, 18.
   Tetrao. 14.
sephæna, Francolinus. 113, 114, 115.
   Perdix. 113.
Severtzov's Hazel-Hen. 77.
Sharpe's Francolin. 116, 128.
sharpii, Francolinus. 116, 128.
Sharp-tailed Grouse. 68.
Sharp-winged Grouse. 57.
shelleyi, Francolinus. 121, 122.
Shelley's Francolin. 121.
Short-tailed Sand-Grouse. 15.
sifanica, Perdix. 151.
Silver Kalij Pheasant. 277.
sinensis, Ithagenes. 219.
Slender-billed Capercailzie. 53.
Smith's Chestnut-vented Sand-Grouse. 19.
Smith's Francolin. 113.
Snow-Cock. 83.
   Altai. 86.
   Caspian. 89.
   Caucasian. 90.
   Himalayan. 86.
   Prince Henry's. 85.
   Tibetan. 84.
Snow Partridges. 79, 80.
Sonnerat's Tree-Partridge. 166.
sonorivox, Bambusicola. 204.
Sooty Capercailzie. 60.
sordidus, Synoicus. 190, 191.
spadicea, Galloperdix. 206.
spadiceus, Francolinus. 206.
   Tetrao. 206.
sphenura, Perdix. 203.
spilogaster, Francolinus. 114.
spilolæmus, Francolinus. 115, 116.
Spitsbergen Ptarmigan. 43.
Spotted Francolin. 114.
Spotted Pin-tailed Sand-Grouse. 14.
Spotted Sand-Grouse. 18.
Spur-Fowl. 205.
   African. 78.
   Ceylon. 210.
   Indian. 78.
   Painted. 208.

Spur-Fowl, Red. 206.
squamatus, Francolinus. 132, 133.
Starna cinerea. 148, 149.
   peregrina. 148.
   rupestris dauurica. 149.
Stone Pheasants. 199.
streptophorus, Francolinus. 112, 113.
striatus, Margaroperdix. 152.
   Perdix. 152.
stuhlmanni, Francolinus. 111.
subtorquatus, Francolinus. 111.
sumatrana, Arboricola. 172.
   Arborophila. 172.
   Caloperdix. 176.
Sumatran Ferruginous Wood-Partridge. 176.
Sumatran Tree-Partridge. 172.
sumatranus, Euplocomus. 244.
superciliaris, Malacortyx. 213.
superciliosa, Ophrysia. 212, 213.
superciliosus, Malacoturnix. 213.
   Rollulus. 213.
swainsoni, Francolinus. 139.
   Perdix. 139.
   Pternistes. 139.
Swainson's Bare-throated Francolin. 139.
Swamp-Quail. 190, 192.
   Raalten's. 192.
Swinhoe's Kalij Pheasant. 278.
swinhoii, Euplocamus. 278.
   Gennæus. 278.
sylvestris, Bonasia. 74.
Syncœcus. 79, 190.
   plumbeus. 192.
   raalteni. 191, 192.
Synoicus. 190.
   australis. 190, 192.
   cervinus. 190.
   diemenensis. 190.
   sordidus. 190, 191.
Syrrhaptes. 3, 7.
   paradoxus. 2, 3, 4, 6, 7, 9.
   tibetanus. 6.
széchenyii, Tetraophasis. 83.
Széchenyi's Pheasant-Grouse. 83.

taigoor, Turnix. 43.
tauricus, Tetraogallus. 89.
temmincki, Ceriornis. 227.
   Tragopan. 227, 230.
temminckii, Satyra. 227.

Temminck's Horned Pheasant. 227.
Tetrao. 49.
  acatoptricus. 48.
  afer. 137.
  albus. 36.
  alchata. 8.
  arenarius. 15.
  betulinus. 74.
  bicalcaratus. 126.
  bonasia. 74.
  canadensis. 54.
  capensis. 129.
  caspius. 89, 90.
  caucasica. 90.
  chinensis. 107, 193.
  coturnix. 180.
  cupido. 62, 65.
  curvirostris. 142.
  dauurica. 149.
  falcipennis. 57.
  franklini. 56.
  gingicus. 166.
  hybridus. 52.
  javanicus. 167.
  kamtschaticus. 54.
  lagopus. 36, 38.
  leucurus. 44.
  madagascariensis. 107, 152.
  manillensis. 196.
  medius. 48.
  mlokosiewiczi. 48.
  mutus. 38.
  namaqua. 11.
  nudicollis. 136.
  obscurus. 58.
  paradoxa. 3.
  parvirostris. 53, 54.
  perdix. 143, 149.
  perlatus. 107.
  petrosus. 97.
  phasianellus. 68.
  pintadeanus. 107.
  pondicerianus. 108.
  porphyrio. 177.
  richardsonii. 61.
  rubricollis. 137.
  rufus. 95.
  rupestris. 42.
  sabinii. 71.
  saliceti. 36.
  scoticus. 27.
  senegallus. 14.

Tetrao spadiceus. 206.
  tetrix. 45.
  togatus. 71.
  umbelloides. 71.
  umbellius. 71.
  uralensis. 52.
  urogalloides. 53.
  urogallus. 49, 52, 53.
  urophasianus. 66.
  viridis. 177.
Tetraogallus. 80, 83.
  altaicus. 86.
  caspius. 89.
  caucasicus. 90.
  challayei. 89.
  henrici. 85.
  himalayensis. 83, 86, 89.
  tauricus. 89.
  tibetanus. 84, 85, 86.
Tetraonidæ. 26.
Tetraophasis. 81.
  desgodinsi. 83.
  obscurus. 81, 82.
  szechenyii. 83.
Tetrastes. 74.
  bonasia. 74, 75.
  griseiventris. 77.
Tetrix uralensis. 50.
tetrix, Lyrurus. 36, 45, 49, 50, 56.
  Tetrao. 44.
textilis, Coturnix. 185.
thoracica, Bambusicola. 202, 203.
  Perdix. 203.
Three-toed Sand-Grouse. 3.
Tibetan Snow-Cock. 84.
Tibetan Three-toed Sand-Grouse. 6.
tibetanum, Crossoptilon. 251, 253.
tibetanus, Crossoptilon. 252.
  Phasianus. 252.
  Syrrhaptes. 6.
  Tetraogallus. 84, 85, 86.
togatus, Tetrao. 71.
torqueola, Arboricola. 160, 165.
  Perdix. 160.
Tragopan. 220.
  blythi. 228.
  caboti. 229.
  hastingsi. 224.
  melanocephalum. 224.
  melanocephalus. 224.
  pucrasia. 284.
  satyra. 220, 224, 225, 227.
  satyrus. 220.

Tragopan temmincki. 227, 230.
Tragopan, Grey-bellied. 228.
Treacher's Tree-Partridge. 170.
Tree-Partridge. 160.
  Aracan. 165.
  Black-throated. 163.
  Brown-breasted. 169.
  Common. 160.
  Formosan. 164.
  Hainan. 164.
  Horsfield's. 171.
  Javan. 167.
  Mandelli's. 167.
  Red-billed. 168.
  Rufous-throated. 165.
  Sonnerat's. 166.
  Sumatran. 172.
  Whitehead's. 171.
tricinctus, Pterocles. 24.
Tringa fasciata. 22.
Tropicoperdix. 161, 172.
  charltoni. 173, 174.
  chloropus. 172.
True Game-Birds. 1, 25.
True Partridges. 143.
Turkeys. 199.
Turnix taigoor. 43.
Tympanuchus. 61.
  americanus. 62, 65, 66.
  cupido. 61, 65.
  pallidicinctus. 65, 66.

Ulu Francolin. 117.
uluensis, Francolinus. 117.
umbelloides, Bonasa. 72.
  Tetrao. 71.
umbellus, Bonasa. 71.
  Tetrao. 71.
umbellus sabini, Bonasa. 71.
umbellus togata, Bonasa. 71.
umbellus umbelloides, Bonasa. 71.
Ural Capercailzie. 52.
uralensis, Tetrix. 50, 52.
urogalloides, Tetrao. 53.

urogallus, Tetrao. 49, 52, 53.
urophasianus, Centrocercus. 66, 67.
  Tetrao. 66.
vaillanti, Perdix. 119.
Variegated Sand-Grouse. 17.
variegatus, Pterocles. 17.
ventralis, Perdix. 199.
Vermicellated Kalij Pheasant. 266, 272.
vieilloti, Euplocomus. 244.
viridis, Tetrao. 177.
vulgaris, Coturnix. 180.
vulgaris japonica, Coturnix. 184.

Wading Birds. 26.
Wattled Pheasant. 248.
  Bulwer's. 249.
welchi, Lagopus. 42.
Western Horned Pheasant. 224.
Western Pin-tailed Sand-Grouse. 8, 10.
White China Pheasant. 277.
White-crested Kalij Pheasant. 258.
Whitehead's Tree-Partridge. 171.
White-tailed Eared-Pheasant. 253.
White-tailed Ptarmigan. 44.
White-throated Francolin. 115.
Willow Grouse. 26, 27, 29, 36, 44, 52, 85.
Wood-Partridge. 172.
  Black. 178, 179.
  Bornean Ferruginous. 176.
  Charlton's. 173.
  Crested. 177.
  Crimson-headed. 174.
  Ferruginous. 175.
  Red-crested. 177, 216.
  Sumatran Ferruginous. 176.

xanthospila, Pucrasia. 285, 286.

Yellow-necked Koklass Pheasant. 286.

zeylonensis, Perdix. 210.

www.ingramcontent.com/pod-product-compliance
Lightning Source LLC
Chambersburg PA
CBHW020245240426
43672CB00006B/641